Current Topics in Microbiology and Immunology

Volume 338

Current Topics in Microbiology and Immunology

Previously published volumes

Further volumes can be found at springer.com

Alan L. Rothman

Editor

Dengue Virus

 Springer

Editor
Dr. Alan L. Rothman
University of Massachusetts
Medical School
Center for Infectious Disease
& Vaccine Research
55 Lake Avenue N.
Worcester MA 01655
USA
alan.rothman@umassmed.edu

ISBN 978-3-642-02214-2   e-ISBN 978-3-642-02215-9
DOI 10.1007/978-3-642-02215-9
Springer Heidelberg Dordrecht London New York

Library of Congress Control Number: 2009929174

Cover motif: A breeding container for Aedes aegypti mosquitoes used for demonstration and teaching, Venezuela (photo: Alan Rothman). Inset-Field: workers check domestic water containers for Aedes aegypti larvae in Kamphaeng Phet, Thailand (photo courtesy of Dr. Thomas Scott).

Cover design: WMXDesign GmbH, Heidelberg, Germany

Printed on acid-free paper

Springer is part of Springer Science+Business Media (www.springer.com)

Preface

Early Scientific Progress

Scientific research on dengue has a long and rich history. The literature has been touched by famous names in medicine – Benjamin Rush, Walter Reed and Albert Sabin, to name a very few – and has been fertile ground for medical historians (Kuno 2009; Rigau-Perez 1998; Kuno 2007; Ashburn and Craig 2004; Pinheiro and Corber 1997; Papaevangelou and Halstead 1977; Halstead 1974; Ehrenkranz et al. 1971). The advances made in those early investigations are all the more remarkable for the limited tools available at the time. The demonstration of a viral etiology for dengue fever (DF), the recognition of mosquitoes as the vector for transmission to humans and the existence of multiple viral variants (serotypes) with only partial cross-protection were all accomplished prior to the ability to culture and characterize the etiologic agent.

Research on dengue in this period was typically driven by circumstances. Epidemics of dengue created public health crises, although these were relatively short-lived in any one location, as the population of susceptible individuals quickly shrank. Military considerations became a major driving force for research. With the introduction of large numbers of nonimmune individuals into endemic areas, dengue could cripple military readiness, taking more soldiers out of action than hostile fire.

Progress on several fronts was significant and engendered optimism that the disease could be controlled. The campaign against *Aedes aegypti* accomplished the elimination of this mosquito from much of the Western hemisphere by 1970. Transmission of DENV and the number of cases were sharply reduced. Working in the Pacific, Dr. Sabin and others isolated dengue viruses (DENV) by serial intracerebral passage in suckling mice and in 1945 reported that the adapted virus was attenuated in humans and could induce protective immunity against the virulent parent virus (Sabin and Schlesinger 1945). Development of an effective vaccine was anticipated to follow.

Unfortunately, by the late 1950s, the picture was beginning to become cloudier. A new disease, dengue hemorrhagic fever (DHF), was recognized in Thailand and the Philippines. The isolation of new DENV serotypes was a focus of early speculation about the pathogenetic basis for this disease but the ability of all DENV serotypes to cause DHF was soon established. Gains in mosquito control in the Western hemisphere were fleeting, as *Aedes aegypti* returned once mosquito control programs were abandoned.

Over the course of the last half of the twentieth century, the global epidemiologic situation worsened, leading some to apply the term "pandemic." Comprehensive data obtained from active surveillance systems were lacking but a steady increase in reported cases of DF and DHF, as well as the number of countries affected, was documented. This steady increase was further punctuated by spikes associated with large epidemics, often involving multiple countries. Although an increase in attention to the disease and a corresponding increased sensitivity for reporting through passive surveillance systems may have contributed to these trends, there is little doubt that the geographic range and intensity of DENV transmission increased. The first occurrences of epidemic DHF in new regions were particularly striking, as in the case of the 1981 epidemic of DHF in Cuba, the first recorded in the Western hemisphere.

Armed with tools for serologic diagnosis and in vitro culture of DENV, scientific progress in understanding and managing dengue disease continued, led by Halstead, Nimannitya, Rosen, Gubler and Bhamarapravati, among others. The importance of plasma leakage as a key feature of DHF facilitated the development of clinical management guidelines that successfully reduced dengue-related morbidity and mortality. Recognition of the predominant infection of monocytic cells, the increased risk for DHF associated with circulation of multiple DENV serotypes and secondary DENV infections and the association of DHF with enhanced cytokine production in vivo guided development of disease models, diagnostic tests and candidate therapeutics. Isolation and in vitro propagation of DENV strains generated an array of viral strains that have been tested as candidate live, attenuated, vaccines.

Despite the overall increase in knowledge during this period, controversies at times overshadowed progress. The early dichotomy between the "viral virulence" and "immune sensitization" models was particularly acute. Accumulated data supported both models. Although some debate continues, a more complex and nuanced picture has evolved incorporating both models and suggesting that human and vector population dynamics, viral evolution and individual and herd immunity all influence the risk for mild versus severe disease.

Dengue at the Start of the Twenty first Century

The Epidemiologic Situation

As the first decade of the twenty first century reaches its end, the epidemiologic trends in dengue disease have given little reason for optimism. Countries that have been endemic for dengue for decades in Southeast Asia, Oceania and the Americas have experienced an increased number of cases (WHO 2007). The number of countries with established endemic DENV transmission has also expanded, with outbreaks of dengue recorded for the first time in Bhutan, Timor-Leste and Nepal (Pandey et al. 2008), the return of dengue to Hawaii after several decades (Effler et al. 2005) and the first outbreaks of DHF recorded in countries that had previously observed only (or predominantly) DF, such as Peru and Brazil (Siqueira et al. 2005). Hyperendemic transmission of all four serotypes has become even further widespread, for example, with the reintroduction of DENV-3 into South America.

With a greater public sensitivity toward emerging infectious diseases, the recent epidemics of dengue have attracted a great deal of attention in the media. Images of tent hospitals being set up in major urban centers, such as in Rio de Janeiro in Brazil in 2008, have been aired on international news programs. The possibility that the global climate will further expand the range of DENV transmission has heightened interest in dengue outbreaks. This increased visibility of dengue as a global health problem has had both positive and negative effects, however. On the positive side, funding for scientific research on dengue from both governmental and nongovernmental sources has surged in the past decade and research papers have been accepted into high-impact, broad-based medical journals. Unfortunately, attention (and resources) has frequently been misdirected at highly visible but ineffective or unproven control strategies (Castle et al. 1999).

Recent Scientific Progress

In contrast to the epidemiologic situation, scientific knowledge on DENV and dengue disease has expanded considerably. Many new investigators have initiated research on DENV and dengue disease during the last decade and new centers for research on tropical or emerging virus diseases have been established by universities and private foundations. These investigators have brought with them new areas of expertise and recently-developed technologies in medicine, virology, molecular biology and immunology. Through their efforts, new insights have been gained into the virion structure, the DENV life cycle, the natural history of infection in humans (and mosquitoes) and the pathogenesis of different manifestations of dengue disease. The remaining chapters of this volume highlight some of these recent advances.

Over the Horizon

In reviewing recent advances, it is fair to ask how these have or will translate into improved global health. Practical applications of recent observations are still theoretical and uncertain. Progress still faces major obstacles, including the lack of a faithful and tractable animal model (although recent developments leave reason for optimism here, as well). The greater depth of understanding of DENV protein structure and function and the complex interactions between the virus and its invertebrate and vertebrate hosts (discussed in the chapters by Paranjape and Harris, Munoz-Jordan, Rico-Hesse and Scott and Morrison) suggest that rational design of effective antiviral drugs may be possible for DENV, as it has been for HIV and HCV (Modis et al. 2003). Combination antiviral and immunomodulatory therapies have generated particular interest (Diamond et al. 2002), given current understanding of dengue disease pathogenesis (discussed in the chapters by Rothman and Stephens). Clinical trials of these drugs will require detailed investigation of viral and host immune response kinetics, based on observations in the natural history of dengue disease (reviewed by Endy and colleagues, Trung and Wills and Srikiatkhachorn and Green).

Most public health professionals would agree that vaccines are likely to be the ultimate solution to control dengue-related morbidity and mortality. As reviewed by Durbin and Whitehead, several of the leading vaccine candidates are the result of advances in molecular biology, using viral strains generated through recombinant DNA technology. Several of the vaccine candidates involve construction of "chimeric" flaviviruses using gene segments from different DENV strains and/or the yellow fever virus 17D vaccine strain. Additional mutations are also being inserted into the DENV genomes in an effort to generate further attenuated strains, as described by Blaney et al.

Summary

Dengue and dengue hemorrhagic fever, which assumed pandemic proportions during the latter half of the last century, have shown no indication of slowing their growth during this first decade of the twenty first century. Challenges remain in understanding the basic mechanisms of viral replication and disease pathogenesis, in clinical management of patients and in control of dengue viral transmission. Nevertheless, new tools and insights have led to major recent scientific advances. As the first candidate vaccines enter large-scale efficacy trials, there is reason to hope that we may soon "turn the corner" on this disease.

Worcester, USA   Alan L. Rothman

References

Ashburn PM, Craig CF (2004) Experimental investigations regarding the etiology of dengue fever. 1907. J Infect Dis 189:1747–1783; discussion 1744–1746
Castle T, Amador M, Rawlins S, Figueroa JP, Reiter P (1999) Absence of impact of aerial malathion treatment on Aedes aegypti during a dengue outbreak in Kingston, Jamaica. Revista panamericana de salud publica = Pan American journal of public health 5:100–105

Diamond MS, Zachariah M, Harris E (2002) Mycophenolic acid inhibits dengue virus infection by preventing replication of viral RNA. Virology 304:211–221

Effler PV, Pang L, Kitsutani P et al. (2005) Dengue fever, Hawaii, 2001-2002. Emerg Infect Dis 11:742–749

Ehrenkranz NJ, Ventura AK, Cuadrado RR, Pond WL, Porter JE (1971) Pandemic dengue in Caribbean countries and the southern United States– past, present and potential problems. N Engl J Med 285:1460–1469

Halstead SB (1974) Etiologies of the experimental dengues of Siler and Simmons. Am J Trop Med Hyg 23:974–982

Kuno G (2007) Research on dengue and dengue-like illness in East Asia and the Western Pacific during the First Half of the 20th century. Rev Med Virol 17:327–341

Kuno G (2009) Emergence of the severe syndrome and mortality associated with dengue and dengue-like illness: historical records (1890–1950) and their compatibility with current hypotheses on the shift of disease manifestation. Clin Microbiol Rev 22:186–201, Table of Contents

Modis Y, Ogata S, Clements D, Harrison SC (2003) A ligand-binding pocket in the dengue virus envelope glycoprotein. Proc Natl Acad Sci USA 100:6986–6991

Pandey BD, Morita K, Khanal SR et al. (2008) Dengue virus, Nepal. Emerging infectious diseases 14:514–515

Papaevangelou G, Halstead SB (1977) Infections with two dengue viruses in Greece in the 20th century. Did dengue hemorrhagic fever occur in the 1928 epidemic? The Journal of tropical medicine and hygiene 80:46–51

Pinheiro FP, Corber SJ (1997) Global situation of dengue and dengue haemorrhagic fever, and its emergence in the Americas. World Health Stat Q 50:161–169

Rigau-Perez JG (1998) The early use of break-bone fever (Quebranta huesos, 1771) and dengue (1801) in Spanish. Am J Trop Med Hyg 59:272–274

Sabin AB, Schlesinger RW (1945) Production of immunity to dengue with virus modified by propagation in mice. Science 101:640–642

Siqueira JB, Jr., Martelli CM, Coelho GE, Simplicio AC, Hatch DL (2005) Dengue and dengue hemorrhagic fever, Brazil, 1981-2002. Emerg Infect Dis 11:48–53

WHO (2007) Report of the Scientific Working Group meeting on Dengue. Geneva, 160

Contents

Contributors

Joseph E. Blaney Jr.
Laboratory of Infectious Diseases, National Institute of Allergy and Infectious Diseases, National Institutes of Health, Bethesda, MD 20892, USA

Anna P. Durbin
Department of International Health, Johns Hopkins Bloomberg School of Public Health
Center for Immunization Research, Department of International Health, Johns Hopkins Bloomberg School of Public Health, Baltimore, MD 21205, USA
adurbin@jhsph.edu

Timothy P. Endy
Associate Professor of Medicine, Chief, Infectious Disease Division, Department of Medicine, State University of New York, Upstate Medical University, 725 Irving Avenue, Suite 304, Syracuse, NY 13210, USA
endyt@upstate.edu

Sharone Green
Center for Infectious Disease and Vaccine Research, University of Massachusetts Medical School, 55 Lake Avenue, North Worcester, MA 01655, USA
Sharone.green@umassmed.edu

Eva Harris
Division of Infectious Diseases, School of Public Health, University of California, Berkeley, 1 Barker Hall, Berkeley, CA 94720-7354, USA

Mammen P. Mammen Jr.
Pharmaceutical Systems, US Army Medical Materiel Development Activity, 1430 Veterans Drive, Fort Detrick, MD 21702, USA
mammen.mammen@us.army.mil

Amy C. Morrison
Department of Entomology, University of California, Davis, CA 95616, USA
amy.aegypti@gmail.com

Jorge L. Muñoz-Jordán
Centers for Disease Control and Prevention, Division of Vector Borne Infectious
Diseases, Dengue Branch, 1324 Calle Cañada, San Juan 00920, Puerto Rico
ckq2@cdc.gov

Brian R. Murphy
Laboratory of Infectious Diseases, National Institute of Allergy and Infectious
Diseases, National Institutes of Health, Bethesda, MD 20892, USA

Suman M. Paranjape
Division of Infectious Diseases, School of Public Health, University of California,
Berkeley, 1 Barker Hall, Berkeley, CA 94720-7354, USA

Rebeca Rico-Hesse
Southwest Foundation for Biomedical Research, San Antonio, TX 78227, USA
rricoh@sfbr.org

Alan L. Rothman
Center for Infectious Disease and Vaccine Research, University of Massachusetts
Medical School, Worcester, MA, USA
alan.rothman@umassmed.edu

Thomas W. Scott
Department of Entomology, University of California, Davis, CA 95616, USA
twscott@ucdavis.edu

Anon Srikiatkhachorn
Center for Infectious Disease and Vaccine Research, University of Massachusetts
Medical School, 55 Lake Avenue, North Worcester, MA 01655, USA
anon.srikiatkhachorn@umassmed.edu

Henry A.F. Stephens
Centre for Nephrology and The Anthony Nolan Trust, University College London,
The Royal Free Hospital Campus, Rowland Hill Street, London, NW3 2PF, UK
h.stephens@ucl.ac.uk

Dinh D Trung
Hospital for Tropical Diseases, Ho Chi Minh City, Vietnam
Department of Infectious Diseases, University of Medicine and Pharmacy of Ho
Chi Minh City, Vietnam
trungdt@oucru.org

Stephen S. Whitehead
Laboratory of Infectious Diseases, National Institute of Allergy and Infectious Diseases, National Institutes of Health, Bethesda, MD 20892, USA
swhitehead@niaid.nih.gov

Bridget Wills
Oxford University Clinical Research Unit, Hospital for Tropical Diseases, Ho Chi Minh City, Vietnam
Paediatric Infectious Diseases, St Mary's Hospital / Imperial College, Praed Street, London, UK
bwills@oucru.org

In-Kyu Yoon
Department of Virology, Armed Forces Research Institute of Medical Sciences, 315/6 Rajvithi Road, Bangkok, 10400 Thailand
InKyu.Yoon@afrims.org

Prospective Cohort Studies of Dengue Viral Transmission and Severity of Disease

Timothy P. Endy, In-Kyu Yoon, and Mammen P Mammen

Contents

Abstract As the four serotypes of dengue virus (DENV) systematically spread throughout the tropical and subtropical regions globally, dengue is increasingly contributing to the overall morbidity and mortality sustained by populations and thereby challenging the health infrastructures of most endemic countries. DENV-human host-mosquito vector interactions are complex and cause in humans either asymptomatic or subclinical DENV infection, mild to severe dengue fever (DF), severe dengue hemorrhagic fever (DHF), or dengue shock syndrome (DSS). Over the past decade, we have seen an increase in research funding and public health efforts to offset the effects of this pandemic. Though multiple vaccine

T.P. Endy (✉)

Infectious Disease Division, Department of Medicine, State University of New York, Upstate Medical University, 725 Irving Avenue, Suite 304, Syracuse, NY 13210, USA
e-mail: endyt@upstate.edu

I.-K. Yoon

Department of Virology, Armed Forces Research Institute of Medical Sciences, 315/6 Rajvithi Road, Bangkok, 10400, Thailand

M.P. Mammen

Pharmaceutical Systems, U.S. Army Medical Materiel Development Activity, 1430 Veterans Drive, Fort Detrick, MD 21702, USA

A.L. Rothman (ed.), *Dengue Virus*, Current Topics in Microbiology and Immunology 338,
DOI 10.1007/978-3-642-02215-9_1, © Springer-Verlag Berlin Heidelberg 2010

development efforts are underway, the need remains to further characterize the determinants of varying severities of clinical outcomes. Several long-term prospective studies on DENV transmission and dengue severity have sought to define the epidemiology and pathogenesis of this disease. Yet, more studies are required to quantify the disease burden on different populations, explore the impact of DENV serotype-specific transmission on host-responses and dengue severity and measure the economic impact of dengue on a population. In this section, we will review the critical past and recent findings of dengue prospective studies on our understanding of the disease and the potential role of future prospective cohort studies in advancing issues required for vaccine field evaluations.

1 Introduction

The global dengue pandemic and its associated morbidity and mortality have drawn in additional research funding to energize the scientific community to pursue a greater understanding of dengue, its virology, pathogenesis, epidemiology and transmission factors. However, there exist only eight published, long-term dengue prospective cohort studies since 1984 (Table 1) that studies disease severity and virus transmission.

Prospective studies offer the advantage of determining the true incidence of disease within a defined cohort to ascertain absolute and relative risk, the spectrum of clinical outcomes (asymptomatic infection to severe hospitalized disease), analysis of risk factors of disease severity and the spatial and temporal diversity of serotype-specific dengue virus (DENV) transmission. Long-term prospective studies can examine the impact of year-to-year variation in dengue incidence, disease severity and serotype-specific transmission. Additionally, prospective-cohort studies and their examination of the full burden of disease can be used to determine the economic burden of DENV infection, essential information for countries in evaluating health priorities and the resources required for vector control and in determining the cost-effectiveness of a DENV vaccine to prevent infection. DENV vaccine developers rely on prospective studies to provide accurate information on dengue incidence for sample-size estimation for efficacy studies and in determining the spatial and temporal transmission of different DENV serotypes for serotype-specific vaccine efficacy. Furthermore, prospective studies develop the field site infrastructure and community awareness necessary to conduct phase III dengue vaccine efficacy studies. The limited number of prospective dengue studies may be a reflection of the associated time and cost necessary for study execution. Additional limitations include: (a) the potential introduction of bias if every member of the cohort is not followed or if the surveillance is limited in identifying all infection or disease, (b) the length of the study may be less than the latency period of the disease such as onset of dengue hemorrhagic fever (DHF) and (c) prospective studies are inherently inefficient for studying rare complications of the disease such as encephalitis.

Table 1 Summary of prospective cohort studies of DENV transmission and disease

Study Site	Population Size[a]	Age Range	Study Period	Dengue Infection	Symptomatic Dengue	Incidence (Average)		
						Hospitalized Dengue	Severe Dengue	Symptomatic: Asymptomatic Ratio
Rayong, Thailand Sangkawibha et al. (1984)	1,056	4–14 years	1980–1981	9.4%	n/a[b]	0.1%	0.3%	n/a
Bangkok, Thailand Burke et al. (1988)	1,757	4–16 years	1980–1981	11.8%	0.7%	0.4%	0.4%	1:8
Yangon, Myanmar Thein et al. (1997)	12,489	1–9 years	1984–1988	5.1%	n/a	0.3%	0.2%	n/a
Yogyakarta, Indonesia Graham et al. (1999)	1,837	4–9 years	1995–1996	29.2%	0.6%	0.4%	0.4%	n/a
Kamphaeng Phet I, Thailand Endy et al. (2002a)	2,119	7–11 years	1998–2002	7.3%	3.9%	1.0%	0.6%	1:0.9
West Java, Indonesia Porter et al. (2005)	2,536	18–66 years	2000–2002	7.4%	1.8%	0.1%	0.1%	1:3
Managua, Nicaragua Balmaseda et al. (2006)	1,186	4–16 years	2001–2002	9.0%	0.85%	n/a	n/a	1:13–1:6
Kamphaeng Phet II, Thailand (14 and unpublished data)	2,095	4–13 years	2004–2006	6.7%	2.2%	0.5%	0.1%	1:3.0

[a]number in cohort tested for dengue antibody (incidence denominator)
[b]n/a = not available; not provided in the published paper

We review in this section the published prospective dengue cohort studies, explore the information acquired from recent studies and discuss research needs for the future as we continue to understand this important disease.

2 Previous Prospective Cohort Studies

Table 1 summarizes the published, prospective dengue cohort studies to date. The first study was conducted in Rayong, Thailand starting in January 1980 among 3,185 children who were randomly sampled from schools and households (Sangkawibha et al. 1984). The population prevalence of neutralizing antibody to the four dengue serotypes was estimated and incidence of infection with each DENV serotype was determined in first grade children who were rebled a year later. Examination of pre- and post-epidemic cohort blood samples revealed that the incidence of dengue infection in 251 seronegative children was 39.4%. Of the 22 shock syndrome cases admitted to the hospital, all had secondary antibody responses based on acute and convalescent serology. The risk factors for dengue shock syndrome (DSS) in Rayong were secondary infections with DENV-2, which followed primary infections with DENV-1, DENV-3 or DENV-4. The Rayong study confirmed the high burden of dengue illness in this population, the association of secondary dengue infections with severe dengue illness and the importance of sequential dengue serotypes in producing shock syndrome.

The second study was a 2 year (1980–1981) school-based study involving 1,757 children, ages 4–16 years, in Bangkok, Thailand (Burke et al. 1988). The children were followed using active surveillance to identify children absent from school for 2 or more consecutive school days. Upon evaluation, if the child had a febrile illness, acute and convalescent serum samples were obtained for serologic testing. Antibody titer revealed that 50% of the enrolled students had evidence of dengue antibody, likely indicative of a DENV infection experienced prior to the age of 7 years. Most (87%) of the students who became infected during the study period were asymptomatic as determined by lack of clinical illness; 53% of the symptomatic DENV infections were recognized as dengue DHF requiring hospitalization. Significant study findings were an incidence in dengue-naïve participants of 6.3%, an incidence of 5.5% in dengue-experienced participants, a hospitalization rate among symptomatic infected children of 53% and a symptomatic-to-asymptomatic ratio of 1:8. The odds ratio for developing DHF in participants with preexisting dengue immunity was ≥ 6.5. This was the first study to determine the full-burden of DENV infection within a cohort and the relationship of preexisting immunity, secondary dengue infection, to dengue disease severity.

A 5 year (1984–1988) study was performed in two townships in Yangon, Myanmar (Thein et al. 1997). A study population of approximately 12,500 children was sampled pre- and post-monsoon season in age-specific cohorts: ages 2, 3, 5 and 6 years. Cohorts in each age group combined varied by year: 1,283 in 1984; 1,513 in 1985; 1,947 in 1986; 1,978 in 1987; and 1,239 in 1988. The total number sampled and tested for dengue antibody was 3,579. Surveillance for severe disease was performed by monitoring hospital admissions at Yangon Children's Hospital. Over 5 years there were 50 hospital admissions diagnosed as dengue fever (DF) (incidence of 0.08% per year) and 145 with DHF or DSS (incidence of 0.02% per year) for a total incidence of symptomatic hospitalized dengue of 0.1% per year. Over 5 years, there were 920 participants who had serologic evidence of DENV infection for an average incidence of 5.1%. The authors concluded that severe DHF/DSS was associated with having serologic evidence of a preexisting DENV infection (secondary dengue).

A 1 year (1995–1996) study was initiated in 1995 in Yogyakarta, Indonesia (Graham et al. 1999). The cohort study involved children ages 4–9 years with blood samples collected at the start of the study and 1 year later. Passive surveillance was performed identifying febrile children presenting to a participating study clinic or admitted to a hospital with suspected dengue. The overall dengue incidence for 1 year was 29.2% with a symptomatic and severe hospitalized dengue illness incidence of 0.6% and 0.4%, respectively.

The Bangkok, Rayong, Yangon and Yogyakarta studies formed the basis of our understanding that secondary DENV infections predispose to risk for subsequent severe dengue upon re-exposure to DENV. Additionally, attempts were made to understand the full burden of DENV infection in a population and the role of serotype-specific DENV transmission. The limitations of these studies were the relatively short periods of observation, mixed ages of cohort populations making incidence determinations difficult and lack of serotype-specific assays to identify the true infecting serotype and subclinical infections. The Bangkok study was the first to determine the full-burden of DENV infection in a well-defined cohort population by using active surveillance to evaluate fevers in children absent from school with acute DENV infections. That study was limited to 1 year and thus variations in serotype exposure and host response from year-to-year could not be determined.

3 Recent Prospective Cohort Studies

Four studies have been carried out in the last decade, one in Nicaragua (the first in the Americas), one in Indonesia (the first in adults) and two in Thailand. The two Thai studies were conducted in sequence for over 7 years of a planned 10 years of continuous observation of dengue disease severity in the province of Kamphaeng Phet. Each study will be discussed in turn, describing the key research advances that have contributed to recent gains in our knowledge of dengue.

A 2 year study was conducted in 1,186 schoolchildren, ages 4–16 years, in Managua, Nicaragua during 2001–2002 (Balmaseda et al. 2006). Blood was drawn in March or May of every year prior to the dengue season. Children who had 3 consecutive days of school absence were evaluated by the school nurse or public health clinic for fever and/or dengue-like illness. The incidence of DENV infection (symptomatic and asymptomatic) in this first cohort study in the Americas was 12% during the first year of the study and 6% during the second year (average incidence of 9%). The incidence of symptomatic dengue was 0.85% and the ratio of symptomatic to asymptomatic infections was 1:13 during the first year of the study and 1:6 during the second year of the study.

A 3 year study was conducted in 2,536 adults, ages 18–66 years, in West Java, Indonesia from 2000– 2002 (Porter et al. 2005). Volunteers were bled every 3 months and actively followed at work for acute illness. The first 2 years of the study demonstrated a symptomatic dengue incidence of 18 per 1,000 person-years and an estimated asymptomatic rate of 56 per 1,000 person-years for a symptomatic to asymptomatic ratio of 1:3.

Two studies, Kamphaeng Phet Study I (KPSI) in 1998–2002 and KPSII in 2004–2008, were performed in Kamphaeng Phet, Thailand. Both studies were conducted in subdistrict Muang, which, as per a year 2000 census, had a population of 198,943 with 49,593 households. Both studies were initiated under a combined National Institutes of Health, United States Army and Thai Ministry of Health-funded grant in collaboration with the University of Massachusetts Medical School, University of California, Davis and the Armed Forces Research Institute of Medical Sciences (AFRIMS). The basis for both Kamphaeng Phet studies was the enrollment of primary school children. Children in KPSI were enrolled as they entered second grade and were followed continuously for up to 5 years or until they graduated at the end of 6th grade. Participating students were evaluated every January with baseline demographic information, height and weight and a blood sample obtained for plasma and peripheral blood mononuclear cells (PBMC's). All participants were evaluated during the first part of June, August and November of each year, when a blood sample was obtained for dengue serology (Endy et al. 2002a; b). In KPSII, children were enrolled from kindergarten to grade 5 and remained in the study for up to a 5 year period; routine blood samples were obtained every June and January to assess for dengue antibody conversion.

For both studies, active acute illness case surveillance of the study participants occurred from June to mid-November, the peak DENV transmission season in Thailand. Acute illness from DENV infection was identified using school absence as the indicator for evaluation. Absent students were identified by their teacher and evaluated by a village health worker with a symptom questionnaire and an oral temperature obtained with a digital thermometer. Students who had a history of fever within 7 days of school absence or an oral temperature $\geq 38^{\circ}$ C (100.4° F) were brought to a public health clinic and evaluated by a nurse. A physical examination was conducted. Acute and convalescent (14 days later) blood samples were obtained. Acutely ill children were also identified if they reported ill to the school nurse or were admitted to the hospital. At the end of each year, based on comparison of dengue antibody responses between blood sampling times in KPSI (June–August, August–November or January–January) and in KPSII (January–January) and evaluation of febrile school absences, acute DENV infections were categorized as: (1) asymptomatic (fourfold rise in dengue antibody titer between sequential blood samples without a reported febrile school absence); (2) symptomatic dengue not requiring hospitalization; (3) symptomatic DF requiring hospitalization; and (4) symptomatic DHF requiring hospitalization. Dengue reverse transcriptase-polymerase chain reaction (RT-PCR) and occasionally, viral isolation, were performed on all acute dengue samples and thus evaluation of the full-burden of DENV infections by virus serotype was attempted. The January blood sampling and PBMC collection provided a valuable archive in which to determine preillness host factors that determine risk for development of severe DENV infection. The results from these studies, summarized below, provide important insights into the dynamics of DENV transmission in a population, the diversity of serotype-specific virus transmission and the host dynamics that produce subclinical to severe dengue illness.

3.1 Dengue Incidence Diversity in Time and Space

For KPS I and II, the overall average annual incidences of DENV infection were 7.3% and 6.7%, respectively; those of symptomatic dengue were 3.9% and 2.2%, those of hospitalized dengue illness 1% and 0.5% and those of severe dengue (DHF) 0.6% and 0.1% respectively. Figure 1 displays the heterogeneity of DENV incidence in a subset of schools that participated in KPS I. In general, dengue incidence was cyclical in each school, with relatively mild years followed by more severe years. Some schools had a severe dengue year, e.g., school 4 in 1998 with an incidence of nearly 20%, while other schools a short distance away had less severe dengue, e.g., school 5 during 1998 with an incidence of less than 5%. The cyclical nature of dengue illness on a temporal and spatial scale is an important concept from these studies and is reflected in the national occurrence of reported dengue illness in Thailand, where relatively mild years are followed by more severe years (Nisalak et al. 2003).

On average, all schools experienced a significant burden of dengue illness during the 4–5 year period (Fig. 1). The importance of understanding dengue incidence temporally and spatially is its utility in understanding the pathogenesis of dengue illness and disease severity. Important questions generated from this information are the role of herd immunity from the previous year's dengue transmission in modifying dengue disease in the subsequent year and how serotype-specific DENV transmission and infection rates are affected. Lastly, dengue incidence and its temporal and spatial diversity in a population is important in designing dengue vaccine efficacy studies and estimating the population and geographic location required to assess statistical efficacy.

3.2 Dengue Serotype and Strain Diversity in Time and Space

In Kamphaeng Phet Province, all four DENV serotypes are known to cocirculate. The spatial and temporal diversity of serotype-specific DENV circulation was not appreciated until the prospective studies were performed. In KPSI, for example, one

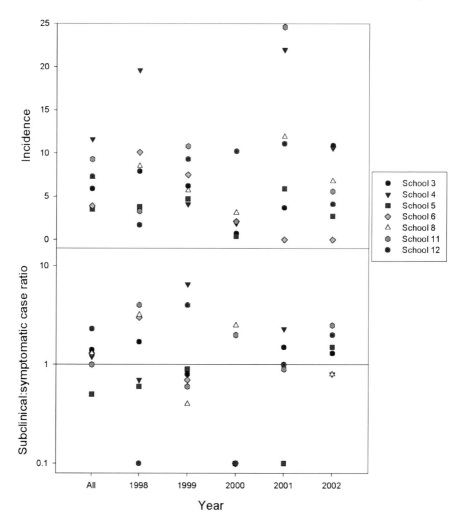

Fig. 1 Variation in incidence of dengue virus infection and ratio of subclinical to symptomatic infections in the Kamphaeng Phet (KPS I) prospective cohort study, 1998–2002. The top figure shows the average incidence for all five years and the incidence for each of the five study years for 7 of the 12 participating schools. The bottom figure shows the ratio of subclinical cases to symptomatic infections for the same seven schools

school had an outbreak with a single DENV serotype while another school a short distance away had a completely different DENV serotype present. Figure 2 demonstrates the spatial and temporal diversity for 1998 through 2001. In 1998, school 4 had a relatively pure DENV-3 outbreak, while a short distance away school 3 had predominantly DENV-1 and school 5 predominantly DENV-2 transmission. The following year, school 4 exhibited predominantly DENV-2 transmission while other schools experienced predominantly DENV-1 transmission. During KPSII

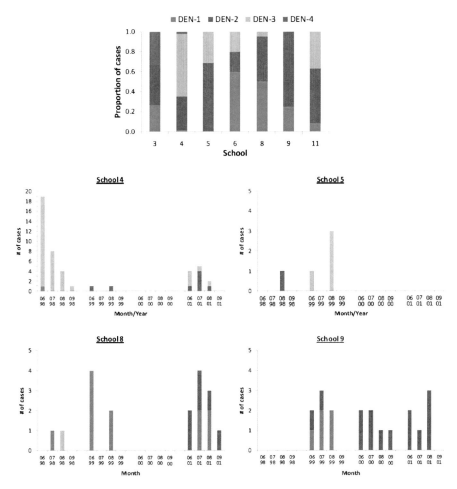

Fig. 2 Spatial and temporal diversity of serotype-specific dengue virus transmission in the Kamphaeng Phet (KPS I) prospective cohort study, 1998–2001. The top figure shows the percentage of infections caused by each dengue viral serotype among all acute infections with serotype information available from 7 of the 12 participating schools during the first 4 years of the study. The bottom four figures show the number of cases of each serotype detected in each of four schools, by month of onset

from 2004– 2007, DENV-1 and DENV-4 were the predominant serotypes. Analysis of isolated viruses from KPSI demonstrated viral genetic variation in both time and space, with multiple viral lineages circulating within individual schools (Jarman et al. 2008). This suggests that there is frequent gene flow of DENV into this microenvironment. Analyses of DENV-2 samples demonstrated clustering of viral isolates within individual schools and evidence of frequent viral gene flow among schools closely related in space. These results suggest that a combination of frequent viral migration into Kamphaeng Phet coupled with population (school) subdivision shape the genetic diversity of DENV at a local scale. Over 5 years of the KPSI study, nearly all schools experienced transmission of three or more DENV serotypes (Fig. 2).

3.3   The Changing Ratio of Subclinical to Clinical Dengue Illness

One value of the prospective studies of dengue disease and virus transmission is to examine the changing relationship between subclinical and clinical DENV infections. Asymptomatic (or subclinical) DENV infection, though a significant component of overall DENV infection, is likely to go unreported despite its potential contribution to the ongoing DENV transmission cycle. Understanding the pathogenetic basis of subclinical infection is important as it reflects a complex interaction between the virus, preexisting DENV antibody and T-lymphocytes and other host factors that determine disease severity and outcome. During KPSI, the changing nature of subclinical to clinical dengue was examined and is illustrated in Fig. 1. In this figure, the ratio of asymptomatic to symptomatic DENV infection was calculated for each school by year. The horizontal line represents a ratio of one, signifying one asymptomatic dengue infection for each case of symptomatic infection. A ratio above this line represents more asymptomatic infections and a ratio below the line represents more symptomatic infections. As illustrated, this ratio was fluid in KPSI, with some schools experiencing more symptomatic disease than other schools and shifting the following year, with others experiencing greater or fewer symptomatic infections. School 4, for example, experienced a ratio of 6.5 during 1999, a ratio near equivalent to what was observed in the prospective cohort study in Bangkok during 1980. The following year this ratio was lower and in 2002 this school experienced a much more severe year than the previous years. Similar patterns were seen for other schools with a cyclic variation in disease severity over time. This suggests that asymptomatic to symptomatic disease ratios at single time points may not reflect the experience of the population over time and that there is an undercurrent of protective immunity that may not prevent infection but may modify disease severity. This was suggested in a study of preillness sera of children who later developed hospitalized dengue; pre-illness heterotypic neutralizing antibody to the child's own DENV isolate was associated with modification of the disease severity (Endy et al. 2004).

3.4   Economic Burden of Dengue Disease

KPSI has provided unique information on the full burden of DENV infection over a 5 year period of time. This information was used to calculate the Disability Adjust Life-Years (DALYs) lost to DENV infections in order to determine the economic impact of dengue in this population (Anderson et al. 2007). The mean cost of dengue was 465.3 DALYs per million population per year, which accounted for

15% of DALYs lost to all febrile illnesses. Non-hospitalized patients with dengue illness represented a substantial proportion of the overall disease burden, 44–73% of the total DALYs lost to dengue each year. The infecting DENV serotype was an important determinant of DALYs lost with DENV-1 responsible for 9% of total DALYs lost, DENV-2 for 30%, DENV-3 for 29% and DENV-4 for 1%. During large outbreak years, DALYs lost to dengue was greater than that calculated for the tropical diseases, meningitis and hepatitis B and three times greater than reported by the World Health Report for 2003 (The world health report 2003). This study demonstrated the under-reporting of DALYs based on reported dengue illness, which focuses on severe hospitalized illness and the value of prospective cohort studies to understand the full economic burden of DENV infection.

3.5   Cluster Investigation

In addition to evaluating the burden of dengue infection, the KPSII study provided unique information on the spatial spread of DENV in the household of dengue-infected children (Mammen et al. 2008). Cluster investigations were conducted within 100 meters of homes where febrile index children with and without DENV infection lived. Information on both human infection and mosquito density and infection were collected to define the spatial and temporal dimensions of DENV transmission. During the first 2 years of the KPSII study, 556 village children were enrolled as neighbors of the index cases. All DENV infections found in these neighbors occurred around DENV-infected index cases, with 12.4% of enrollees becoming infected in a 15 day period. This study demonstrated the focal nature of DENV transmission and the value of cluster investigations as an adjunct to prospective cohort studies in determining the full scope of DENV transmission.

4   Summary

Despite the diversity in the study techniques and populations studied, several recurrent themes emerge from the published prospective dengue cohort studies. First, the incidence of DENV infection in the countries studied is significant, with a range of 5%–29% per year with most studies establishing an annual incidence rate between 5 and 10%. Second, symptomatic DENV infections represent only a fraction of the full burden of DENV infection with incidence ranging from 0.6% to 4% per year. Third, the ratio of symptomatic to asymptomatic DENV infection ranges from 1:1 to 1:8, suggesting variability in subclinical infections contributing to the overall dengue burden. This variability, however, may be attributed to differences in surveillance approaches, prior DENV exposure and/or host genetics. From the Kamphaeng Phet studies, this variability was observed even within a small geographic area over time and was dependent on heterotypic protective immunity from prior DENV exposure and the varying predominance of circulating DENV serotypes each year. Lastly, important information was gathered to evaluate the economic burden of dengue on a population and the potential cost-effectiveness of a vaccine in alleviating morbidity and the burden on the health infrastructure.

Prospective cohort dengue studies provide invaluable reagents (pre-illness sera and peripheral blood mononuclear cells) for further scientific discovery of pathogenetic determinants. Critical questions remain regarding the mediators of severe dengue and the correlates of protection. Additional research is needed to address these areas that are critical to vaccine testing and evaluation. Expanding cohort studies, to include those countries (especially in the Americas and South Asia) where dengue is inadequately characterized, will enable us to further understand the unique host factors that may contribute to differences in dengue risk and may thereby underlie potential differences in vaccine response. Investing in long-term follow-up of cohort populations is important to understand the spatial and temporal variability in DENV circulation and the impact of preexisting dengue antibody on dengue disease severity.

The prospective cohort study is an important method to understand the full-burden of DENV infection, the effects of serotype-specific DENV transmission and the effects of the virus-host interactions that result in mild to severe dengue illness.

References

Anderson KB, Chunsuttiwat S, Nisalak A, Mammen MP, Libraty DH, Rothman AL et al (2007) Burden of symptomatic dengue infection in children at primary school in Thailand: a prospective study. Lancet 369(9571):1452–1459

Balmaseda A, Hammond SN, Tellez Y, Imhoff L, Rodriguez Y, Saborio SI et al (2006) High seroprevalence of antibodies against dengue virus in a prospective study of schoolchildren in Managua, Nicaragua. Trop Med Int Health 11(6):935–942

Burke DS, Nisalak A, Johnson DE, Scott RM (1988) A prospective study of dengue infections in Bangkok. Am J Trop Med Hyg 38(1):172–180

Endy TP, Chunsuttiwat S, Nisalak A, Libraty DH, Green S, Rothman AL et al (2002a) Epidemiology of inapparent and symptomatic acute dengue virus infection: a prospective study of primary school children in Kamphaeng Phet, Thailand. Am J Epidemiol 156(1):40–51

Endy TP, Nisalak A, Chunsuttiwat S, Libraty DH, Green S, Rothman AL et al (2002b) Spatial and temporal circulation of dengue virus serotypes: a prospective study of primary school children in Kamphaeng Phet, Thailand. Am J Epidemiol 156(1):52–59

Endy TP, Nisalak A, Chunsuttitwat S, Vaughn DW, Green S, Ennis FA et al (2004) Relationship of preexisting dengue virus (DV) neutralizing antibody levels to viremia and severity of disease in a prospective cohort study of DV infection in Thailand. J Infect Dis 189(6):990–1000

Graham RR, Juffrie M, Tan R, Hayes CG, Laksono I, Ma'roef C et al (1999) A prospective seroepidemiologic study on dengue in children four to nine years of age in Yogyakarta, Indonesia I. studies in 1995–1996. Am J Trop Med Hyg 61(3):412–419

Jarman RG, Holmes EC, Rodpradit P, Klungthong C, Gibbons RV, Nisalak A et al (2008) Microevolution of dengue viruses circulating among primary school children in Kamphaeng Phet, Thailand. J Virol 82(11):5494–5500

Mammen MP, Pimgate C, Koenraadt CJ, Rothman AL, Aldstadt J, Nisalak A et al (2008) Spatial and temporal clustering of dengue virus transmission in Thai villages. PLoS Med 5(11):e205

Nisalak A, Endy TP, Nimmannitya S, Kalayanarooj S, Thisayakorn U, Scott RM et al (2003) Serotype-specific dengue virus circulation and dengue disease in Bangkok, Thailand from 1973 to 1999. Am J Trop Med Hyg 68(2):191–202

Porter KR, Beckett CG, Kosasih H, Tan RI, Alisjahbana B, Rudiman PI et al (2005) Epidemiology of dengue and dengue hemorrhagic fever in a cohort of adults living in Bandung, West Java, Indonesia. Am J Trop Med Hyg 72(1):60–66

Sangkawibha N, Rojanasuphot S, Ahandrik S, Viriyapongse S, Jatanasen S, Salitul V et al (1984) Risk factors in dengue shock syndrome: a prospective epidemiologic study in Rayong, Thailand. I. The 1980 outbreak. Am J Epidemiol 120:653–669

The world health report (2003). Annex table 3: burden of disease in DALYs by cause, sex and mortality stratum in WHO Regions, estimates for 2002. Geneva: World Health Organization; 2004.

Thein S, Aung MM, Shwe TN, Aye M, Zaw A, Aye K et al (1997) Risk factors in dengue shock syndrome. Am J Trop Med Hyg 56(5):566–572

Control of Dengue Virus Translation and Replication

Suman M. Paranjape and Eva Harris

Contents

S.M. Paranjape
Science and Technology Policy Fellowship Program, American Association for the Advancement of Sciences, 1200 New York Avenue NW, Washington DC, USA 20005
e-mail: sumip1@gmail.com

E. Harris (✉)
Division of Infectious Diseases, School of Public Health, University of California, 1 Barker Hall, Berkeley, DC, CA 94720-7354, USA
e-mail: eharris@berkeley.edu

A.L. Rothman (ed.), *Dengue Virus*, Current Topics in Microbiology and Immunology 338,   15
DOI 10.1007/978-3-642-02215-9_2, © Springer-Verlag Berlin Heidelberg 2010

Abstract Dengue poses an increasing threat to public health worldwide. Studies conducted over the past several decades have improved our knowledge of the mechanisms of dengue virus translation and replication. New methodologies have facilitated advances in our understanding of the RNA elements and viral and host factors that modulate dengue virus replication and translation. This review integrates research findings and explores future directions for research into the cellular and molecular mechanisms of dengue virus infection. Lessons learned from dengue virus will inform approaches to other viruses and expand our understanding of the ways in which viruses co-opt host cells during the course of infection. In addition, knowledge about the molecular mechanisms of dengue virus translation and replication and the role of host cell factors in these processes will facilitate development of antiviral strategies.

1 Introduction

1.1 Dengue Virus Background

Dengue virus (DENV) poses a rapidly increasing threat to public health worldwide. A member of the family *Flaviviridae* and the genus *Flavivirus*, DENV is an enveloped, positive-sense RNA virus with a ~10.7 kb genome that exists as four serotypes (DENV1–4) and is related to other flaviviruses including West Nile, Japanese encephalitis and yellow fever viruses. DENV is the causative agent of dengue fever and the potentially lethal dengue hemorrhagic fever and dengue shock syndrome and is transmitted to humans by the bite of infected *Aedes aegypti* and *Ae. albopictus* mosquitoes. The increasing global spread of DENV and the lack of an approved vaccine or anti-viral has prompted intensive research over the past several years. While the DENV field has benefitted from concurrent research on other flaviviruses (Clyde et al. 2006; Lindenbach and Rice 2003; Westaway et al. 2002), differences between the viruses underscore important mechanistic variations. This review focuses on recent advances in understanding DENV translation and replication, mechanisms that are central to virus propagation and important targets for pharmaceutical development.

1.2 DENV Lifecycle

DENV enters the host cell via receptor-mediated endocytosis and, following pH-mediated fusion of the DENV envelope with the endosomal membrane, undergoes nucleocapsid escape into the cytoplasm. The positive-sense RNA genome is transported to the endoplasmic reticulum (ER), where it is translated to produce viral proteins, including the nonstructural (NS) proteins that are required for subsequent rounds of RNA synthesis. The minus-strand DENV RNA is synthesized and serves as a template for replication of genomic positive-sense RNA molecules, which then undergo translation, generating structural proteins required for RNA packaging and virus assembly. This coordinated process yields virions which, following maturation in the Golgi apparatus, are secreted via the host secretory pathway.

1.3   Conserved Sequences and Structures in the DENV 5′ and 3′ Untranslated Regions

Conserved sequences and structures at the 5′ and 3′ ends of the DENV genome (Fig. 1) are critical for regulation of translation and replication. In DENV and other

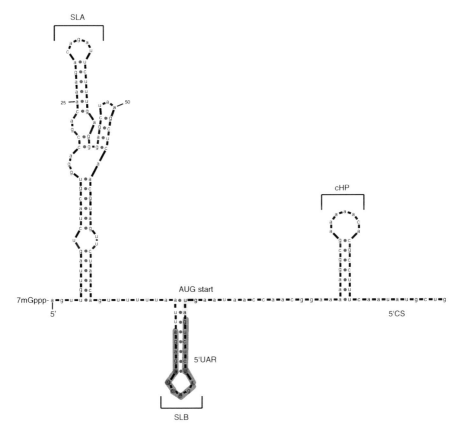

Fig. 1 (a) Conserved sequences and structures in the DENV untranslated regions (UTRs). The DENV 5′UTR harbors three structured regions and two conserved sequences. The 5′ conserved sequence (5′CS) and 5′ upstream UAG region (5′UAR), with complementary sequences in the 3′UTR, are highlighted in yellow and pink, respectively. The capsid initiation codon, AUG, is highlighted in green. Stem-loop A (SLA), SLB and the capsid hairpin (cHP) structure are indicated. Sequence corresponds to DENV2 strain 16681

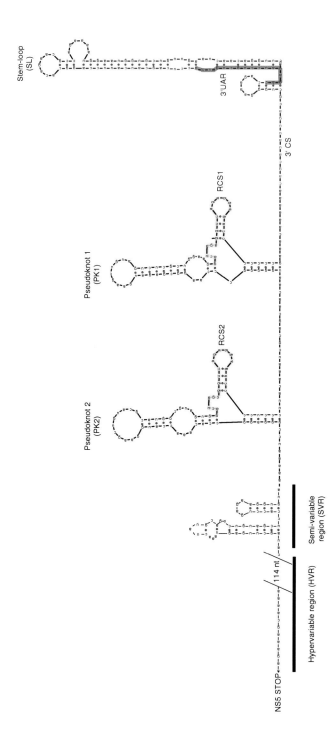

Fig. 1 (b) Conserved sequences and structures in the DENV untranslated regions (UTRs). The predicted structure of the DENV 3′UTR contains three structured elements and four highly conserved sequence elements. The variable region (**VR**) comprises a hypervariable region (HVR) and a semi-variable (SVR) region. Downstream of this are two pseudoknot (PK) structures, PK2 and PK1. The side stem-loops of these structures contain repeated conserved sequence elements (RCS2 and CS2, in green). Located at the terminal 3′ end of the DENV genome is a highly stable 3′ stem-loop structure (3′SL). Adjacent to the 3′SL are the 3′CS (yellow) and the 3′UAR (orange), which are complementary to the 5′CS and 5′UAR sequences, respectively. Sequence corresponds to DENV2 strain 16681

flaviviruses, the 5′ conserved sequence (CS), located downstream of the capsid AUG and the 5′ upstream-AUG-region (UAR) display complementary to the 3′CS and 3′UAR, respectively, in the 3′untranslated region (UTR) (Hahn et al. 1987; Shurtleff et al. 2001). It has been demonstrated using atomic force microscopy that the 5′ and 3′CS and the 5′ and 3′UAR of DENV physically anneal to mediate genome cyclization (Alvarez et al. 2005b), while functional analysis indicates that CS and UAR complementarity is required for DENV replication and RNA synthesis (Alvarez et al. 2008; Alvarez et al. 2005a; Khromykh et al. 2001a; Khromykh et al. 2000; Markoff 2003; You et al. 2001).

Despite divergence of primary sequence, several predicted secondary structures within the 5′ and 3′ ends of flavivirus genomes are maintained (Brinton and Dispoto 1988; Mohan and Padmanabhan 1991; Shurtleff et al. 2001). Secondary structure predictions indicate the presence of two stem-loops (SL), SLA and SLB, within the 98–100 nt 5′UTR of DENV1–4 and another conserved stem-loop, the capsid hairpin (cHP), immediately downstream of the 5′UTR within the capsid coding region (Clyde and Harris 2006). Analysis of the solution structure of the DENV2 5′UTR confirms the existence of each of these secondary structures (Polacek et al. 2009a). Two conserved pseudoknot (PK) structures, PK1 and PK2 and the terminal 3′ stem-loop (3′SL) are predicted by the sequence of the 400–488 nt 3′UTR of DENV1–4 as well as other flaviviruses. Analysis of the solution structure of the terminal 100 nucleotides confirms the stability of the 3′SL (Mohan and Padmanabhan 1991). In addition, the penta-nucleotide loop sequence within the 3′SL and repeated conserved sequences (CS2 and RCS2) within PK1 and PK2, respectively, display strong sequence conservation across DENV serotypes and flavivirus genotypes. Maintenance of these sequences and structures throughout flavivirus evolution indicates their critical involvement in the viral lifecycle.

2   Translation of DENV

2.1   Mechanism of DENV Translation

DENV, like all viruses, utilizes the host cell machinery for translation. DENV translation produces a single polyprotein that is co- and post-translationally cleaved by viral and host proteases to yield 3 structural proteins (capsid, membrane and envelope) and 7 nonstructural proteins (NS1, NS2A, NS2B, NS3, NS4A, NS4B and NS5). The primary mechanism of DENV translation occurs via cap-dependent initiation (Chiu et al. 2005; Holden and Harris 2004), in which the DENV 5′ type 1 7-methylguanosine cap (Cleaves and Dubin 1979), conferred by RNA cap methylation activity of NS5 (Dong et al. 2007), is recognized and bound by eIF4E. eIF4E, in turn, recruits the scaffolding protein eIF4G and the helicase eIF4A. eIF3 association with this complex mediates binding of the 43S ribosomal pre-initiation complex, which, after scanning and identification of the initiation codon, associates with the 60S subunit to form an 80S ribosomal complex that is competent for elongation. The efficiency of initiation and ribosome recycling is enhanced in nearly all cellular mRNAs by a 3′-poly(A) tail. The poly(A) tail functions to recruit poly(A)-binding protein (PABP), which, in turn, binds to eIF4G, circularizing the mRNA and facilitating initiation and/or ribosome recruit-ment. Although the DENV 3′UTR lacks poly(A) sequences, it nonetheless enhances DENV translation (Chiu et al. 2005; Holden and Harris 2004), possibly through a circularization mechanism similar to other mRNAs (Edgil and Harris 2006). Mechanisms to enhance DENV translation are likely important because, although many viruses have evolved mechanisms to shut off host cell translation, DENV does not disrupt host cell protein synthesis. Instead, DENV appears to overcome competition for host cell machinery via subcellular localization of viral processes to ER-derived membrane structures, ensuring that replication, translation and packaging are efficiently orchestrated in potentially isolated replication foci (Mackenzie et al. 1998; Miller et al. 2007; Miller et al. 2006). In addition, DENV may also gain a translational advantage in host cells via a noncanonical mechanism that enables DENV translation to occur under adverse cellular conditions that inhibit cellular cap-dependent translation (Edgil et al. 2006).

2.2   Cis *Elements that Regulate DENV Translation*

Translation is a pivotal step in the DENV lifecycle and deficiencies in translation can significantly reduce viral replication. Supporting this, phosphorodiamidate morpholino oligos (PMOs) targeted to the 5′SLA significantly reduced translation of DENV replicons as well as viral replication (Holden et al. 2006; Kinney et al. 2005). It appears that 5′CS sequences are not involved in DENV translation. Interestingly, substitution of the DENV 5′UTR with EMCV IRES or 5′beta-globin sequences results in wild-type levels of translation (Chiu et al. 2005; Holden and Harris 2004), suggesting functional similarities and the likelihood that similar host factors mediate viral and cellular translation. In addition to regulation at the stage of initiation, mechanisms exist to ensure correct start site selection in both mammalian and mosquito cells. The DENV 5′cHP functions in a position-dependent, sequence-independent manner to overcome the poor initiation context of the first AUG and enhance its selection as the start codon of the capsid gene (Clyde and Harris 2006). The presence of alternative, in-frame AUGs downstream of the cHP within capsid raises the question of whether alternate translation products are involved in regulation of DENV replication.

Although the DENV 3′UTR lacks a poly(A) tail that is the hallmark of nearly all cellular mRNAs, it nonetheless plays an important role in regulation of translation efficiency. For instance, the low replicative capacity of certain Nicaraguan DENV2 strains compared to Asian strains was due, at least in part, to mutations in the 3′UTR that reduce levels of translation in *in vitro* extracts and in cultured cells (Edgil et al. 2003). Deletion mutagenesis in the context of a reporter RNA and a DENV2 replicon indicated that PK1 and PK2 domains had a minor impact on translation (Alvarez et al. 2005a; Chiu et al. 2005), whereas deletion of the 3′SL and the adjacent 3′SLB reduced translation of reporter RNAs by at least 50% (Chiu et al. 2005; Holden and Harris 2004). Furthermore, PMOs that disrupt the pentanucleotide loop at the top of the 3′SL also reduced input translation of a DENV2 replicon by 50% (Holden et al. 2006). As expected, deletion of the 3′UTR significantly reduced translation of DENV reporter constructs. Replacement of the DENV 3′UTR with a poly(A) tail restored translation to wild-type levels (Chiu et al. 2005; Holden and Harris 2004), suggesting that poly(A) binding protein (PABP) may regulate DENV translation. In contrast, deletion of the entire 3′UTR in a DENV2 replicon had no impact on input translation (Alvarez et al. 2005a), suggesting a possible role for heretofore unidentified translational enhancers in the coding region or that subcellular localization of replicon RNA during translation might lead to altered requirements for DENV translation.

2.3 Non-canonical Mechanism of DENV Translation

In addition to canonical cap-dependent translation, a novel mechanism has been described wherein DENV translation and replication occur under conditions such as high osmolarity, reduced eIF4E, or the absence of a functional 5′cap that severely inhibit cellular translation (Edgil and Harris 2006; Edgil et al. 2006). This noncanonical translation requires eIF4G and occurs via a 5′ end-dependent, internal ribosome entry site (IRES)-independent mechanism (Edgil et al. 2006; Paranjape, Polacek, and Harris, unpublished). Further characterization of this noncanonical mechanism will enable insight into DENV replication under adverse cellular conditions.

2.4 Identification of Host Factors that Regulate DENV Translation

Just as host-derived basal translation factors are required for translation initiation and elongation, it is likely that other host factors interact with DENV cis-elements to regulate translation and determine the extent of infectivity within a cell. To this end, several mammalian proteins, including eukaryotic initiation factor 1A (eIF1A), poly-pyrimidine tract binding protein (PTB), Y-box binding protein 1 (YB-1) and heterogenous nuclear ribonucleoproteins (hnRNP A1, hnRNP A2/B1 and hnRNP Q) have been demonstrated to bind to the DENV 3′UTR (Blackwell and Brinton 1995; Blackwell and Brinton 1997; Davis et al. 2007; De Nova-Ocampo et al. 2002; Paranjape and Harris 2007; Yocupicio-Monroy et al. 2007; Yocupicio-Monroy et al. 2003) (Fig. 2). Mosquito and human La autoantigen (La) have been

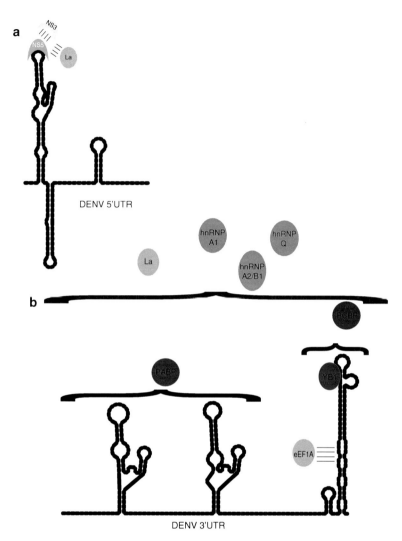

Fig. 2 Host and viral proteins interact with the DENV UTRs. Defined interactions are depicted by direct association of factors with the RNA, while hypothesized interactions are depicted by dashed lines. Factors designated in blue have a demonstrated or postulated role in RNA synthesis while those depicted in red have been shown or are hypothesized to regulate DENV translation. (**a**) Host and viral proteins that interact with the DENV 5′UTR. NS5 interacts with the loop region of the 5′SLA and can interact with NS3. La interacts with the 5′UTR of DENV and with viral proteins NS3 and NS5, although it is not clear whether these interactions occur simultaneously. (**b**) Host and viral proteins that bind the DENV 3′UTR. YB-1 associates with the loop region of the 3′SL. Dashed lines indicate presumptive binding of eEF1A to bulges in the DENV 3′SL. PABP binds to the 3′UTR upstream of the 3′SL. La and the hnRNPs A1, A2/B1 and Q bind to undetermined regions in the DENV 3′UTR

shown to interact with both the DENV 5′ and 3′ UTRs (Garcia-Montalvo et al. 2004; Yocupicio-Monroy et al. 2007; Yocupicio-Monroy et al. 2003). Recent work has demonstrated that PABP binds the 3′UTR upstream of the 3′SL (Polacek et al. 2009b), perhaps explaining the functional redundancy of the DENV 3′UTR and a poly(A) tail in translational enhancement.

Although the DENV UTR-interacting factors mentioned above are involved in host cell processes including splicing, transcription, RNA trafficking and translation, virtually nothing is known about their roles in the DENV lifecycle. One exception is recent work indicating that YB-1 binds to the terminal loop of the 3′SL and has an anti-viral effect on DENV infection. Using genetically deficient mouse embryo fibroblasts, it was determined that this anti-viral impact was mediated at least in part by a repressive effect on input strand translation. YB-1-mediated translational repression is not observed with reporter constructs *in vitro* or in cultured cells (Paranjape and Harris 2007) and is not affected by DENV proteins supplied in *trans*, indicating that translational repression requires genomic sequences and/or viral proteins provided in *cis*. Further experiments will clarify the mechanism of other DENV UTR-associating host factors in viral translation and may help explain tropism at a molecular level.

3 Replication of DENV

3.1 Mechanism of DENV Replication

RNA synthesis is critical for viral propagation and is a predominant target for development of antiviral therapeutics and attenuated vaccine strains (Blaney et al. 2004; Blaney et al. 2008; Butrapet et al. 2000; Keller et al. 2006; Patkar and Kuhn 2006). Immunolocalization studies of DENV and other flaviviral replication complexes (RCs) have demonstrated that viral replication occurs in close association with induced cellular membranes derived from the ER (Mackenzie et al. 1996; Westaway et al. 1999; Westaway et al. 1997). The initial step of viral replication involves the *de novo* synthesis of a negative-strand intermediate using the positive-sense genomic strand as a template. The resultant double-stranded RNA (dsRNA) intermediate, the replicative form (RF), subsequently serves as the template for the synthesis of the plus-strand DENV genomic RNA via a replicative intermediate complex (RI) (Fig. 3). Viral replication proceeds on this RI asymmetrically and semi-conservatively, resulting in predominant production of positive- versus nega-tive-strand RNA (Khromykh and Westaway 1997). Biochemical analyses indicated that approximately 5 nascent strands are present on each RI and that synthesis of each strand required approximately 12–15 mins for completion (Cleaves et al. 1981). Dynamic interaction of cis-RNA elements and subcellular localization of virus-derived replication components enables orchestration of DENV genomic RNA synthesis, translation of viral proteins and packaging.

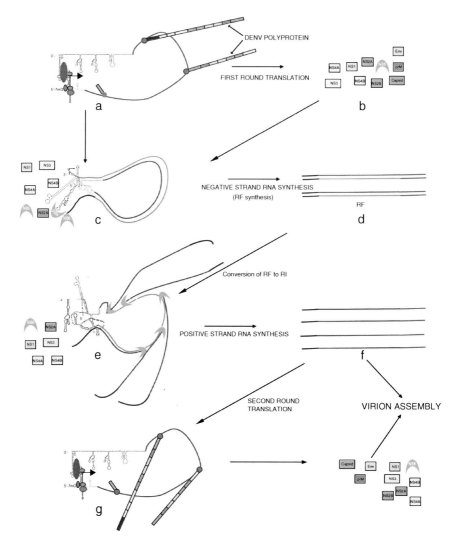

Fig. 3 Molecular regulation of DENV replication relies upon *cis*- and *trans*-elements to modulate genome circularization and facilitate translation and RNA synthesis. (**a**) Upon endosomal escape, translation of the positive-sense DENV RNA is mediated by host-cell factors. One mechanism of translational regulation is likely to involve host protein-mediated genome circularization. (**b**) Translation yields a single polyprotein that is cleaved co- and post-translationally to yield DENV nonstructural proteins that are required for replicase formation and function. (**c**) Minus-strand RNA synthesis involves genome circularization mediated by interactions of the 5′CS and 5′UAR with the 3′CS and 3′UAR, respectively and association of NS5 with the 5′SLA. Presumably, minus-strand synthesis begins upon assembly of NS5 with the other components of the DENV replicase, NS3, NS1, NS2A, NS4A and NS4B. (**d**) Minus-strand synthesis results in formation of the double-stranded RF. (**e**) By mechanisms that are not well-understood, the RF undergoes transition to a replicative intermediate (RI), a process that likely also involves genome cyclization. (**f**) Large amounts of positive-strand RNA are generated using the minus-strand

3.2   Cis-*Elements that Regulate DENV RNA Synthesis*

Efficient RNA synthesis, which is imperative for DENV propagation, is regulated by conserved sequences and structures in the 5′ and 3′ UTRs. In addition to binding essential components of the viral replicase, *cis*-elements mediate the formation of dynamic secondary and tertiary structures that likely facilitate replication and transitions between replication and translation. Functional evidence indicates that genome cyclization, mediated by complementary 5′/3′ CS and 5′/3′ UAR sequences (Fig. 3), is required for negative-strand RNA synthesis of DENV and other flaviviruses (Alvarez et al. 2005a; Alvarez et al. 2008; Alvarez et al. 2006; Alvarez et al. 2005b; Filomatori et al. 2006; Hahn et al. 1987; Lo et al. 2003; You et al. 2001; Yu et al. 2008; Zhang et al. 2008). Importantly, deletions within the stem or loop regions of the 5′SLA in DENV2 infectious clones resulted in decreased RNA synthesis and DENV replication (Filomatori et al. 2006), likely by abolishing interaction with the RNA-dependent RNA polymerase (RdRp), NS5, which has been shown to bind to the SLA *in vitro* (Filomatori et al. 2006)(see below). Moreover, mutagenesis of SLB indicated that, while maintenance of the 5′UAR sequence was important, the structural features of the element were not required (Alvarez et al. 2008). Deletion of conserved nucleotides 82–87 within the 5′UTR SLB of the DENV4 infectious clone had a moderate impact upon RNA synthesis and viral production in mammalian $LLCMK_2$ cells but abolished RNA and virion production in C6/36 cells and *Aedes* mosquitoes (Cahour et al. 1995). Interestingly, mutations in the cHP have recently been shown to abolish RNA synthesis (Clyde et al. 2008), indicating a requirement for this element in translation and RNA synthesis.

Sequences and structures within the DENV 3′UTR play a pivotal role in regulating negative strand RNA synthesis, which initiates at the 3′ terminus of the genome. At the 5′ end of the 3′UTR, deletion of the variable region (VR) in a DENV2 replicon reduced RNA synthesis in BHK cells but had no impact on RNA synthesis in C6/36 mosquito cells (Alvarez et al. 2005a) and VR mutations in a DENV1 infectious clone decreased RNA accumulation in Vero cells (Tajima et al. 2007). Detailed analysis of the regions within the DENV1 VR indicated that the length of the hypervariable region (HVR), rather than its sequence, was important for viral RNA synthesis whereas, conversely, specific sequences of the semi-variable region (SVR) were required for wild-type levels of RNA synthesis (Tajima et al. 2007). Moving downstream, individual deletions of either PK1 or PK2 in the context of a DENV2 replicon resulted in decreased RNA replication, while deletion of both elements completely abolished RNA synthesis, suggesting that these elements enhance DENV RNA synthesis (Alvarez et al. 2005a). Interestingly, deletion of a 30-nt region (Δ30) that comprises the long stem-loop of PK1 resulted in decreased RNA synthesis of DENV2 replicons in BHK and C6/36 mosquito cells (Alvarez

←——

Fig. 3 (Continued) as a template. (**g**) Resultant positive-sense genomes undergo the second round of translation, producing structural proteins for virion assembly. Replicated and translated RNA molecules are packaged with DENV structural proteins and secreted

et al. 2005a), supporting the idea that reduction of RNA synthesis contributes to attenuation of DENV1 and DENV4 vaccine strains that harbor the same deletion (Durbin et al. 2006; Durbin et al. 2005; Whitehead et al. 2003). At the very 3′ terminus of the DENV genome, the highly stable and conserved DENV 3′SL plays a pivotal role in viral RNA synthesis. Mutations that altered the secondary structure of the top half of the DENV2 3′SL drastically reduced negative-strand RNA synthesis (Zeng et al. 1998). Furthermore, PMOs targeted to the highly conserved pentanucleotide loop region of the 3′SL led to a dramatic reduction in RNA synthesis of a DENV2 replicon (Holden et al. 2006), consistent with the strict requirement for this sequence in WNV RNA replication (Tilgner et al. 2005). Mutagenic analysis indicates that the formation of pseudoknot structures between the bottom of the 3′SL stem and the adjacent, smaller stem-loop is important for negative-strand RNA synthesis *in vitro* (You et al. 2001). Finally, *in vitro* analyses indicated that the 3′ terminal nucleotide, U, of DENV was required for the initiation of negative-strand RNA synthesis (Nomaguchi et al. 2003) although, in contrast to WNV replication (Khromykh et al. 2003; Nomaguchi et al. 2003), the conserved penultimate C is not required.

While it is clear that conserved sequences and structures of DENV play an important role in regulating viral RNA synthesis, our understanding of the requirements for positive-strand RNA synthesis are rudimentary. The development of *in vitro* tools will allow delineation of the roles of *cis* RNA elements and *trans* host and viral factors in conversion of the RF to RI and subsequent production of positive-strand progeny.

3.3   Viral Proteins Involved in DENV Replication

DENV nonstructural proteins comprise a viral replicase that includes NS1, NS2A, NS3, NS4A, NS4B and NS5 (Fig. 3). NS5, a multidomain protein harboring RNA cap methylation and RdRp activities (Egloff et al. 2002; Selisko et al. 2006), is required for negative- and positive-strand viral RNA synthesis (Ackermann and Padmanabhan 2001; Bartholomeusz and Wright 1993; Nomaguchi et al. 2003; Raviprakash et al. 1998; Tan et al. 1996). Mutation of the conserved polymerase motif, GDD, to nonfunctional GVD abolished replication of infectious clone and replicon RNA upon transfection (Alvarez et al. 2005a; Holden et al. 2006). Recently, an important step of template discrimination has been shown to be achieved by binding of DENV NS5 to the loop region of the 5′SLA (Fig. 2); viral genome circularization via complementary sequences at the 5′ and 3′ ends of the DENV genome presumably ensures that NS5 bound at the 5′SLA is delivered to the 3′UTR for synthesis of negative-strand RNA (Filomatori et al. 2006). Interestingly, NS5 was demonstrated to be differentially phosphorylated during infection, with phospho-NS5 accumulating in the nucleus of infected cells (Kapoor et al. 1995). Modulation of NS5 association with DENV RNA may play an important role in the transition from translation to replication and in the conversion of the RF to the RI.

Another essential component of the replicase is the multi-functional DENV NS3 protein, which possesses an N-terminal serine protease domain, a DExII-subfamily helicase characterized by seven helicase motifs, and a polynucleotide-stimulated NTPase domain. Mutagenesis of NS3 demonstrated that the helicase and NTPase domains were required for viral replication (Matusan et al. 2001). DENV NS3 and hypo-phosphorylated NS5 were shown to interact in cultured cells (Kapoor et al. 1995) and the two proteins were shown to be essential for RNA synthesis (Cui et al. 1998) and conversion of the RF to RI form *in vitro* (Ackermann and Padmanabhan 2001; Bartholomeusz and Wright 1993; Raviprakash et al. 1998). The crystallization of NS3 (Xu et al. 2006) has provided insight into the molecular regulation of this molecule and has stimulated development of targeted helicase inhibitors (Sampath et al. 2006).

Although NS1, NS4A and NS4B colocalize with the replicase by immunolocalization studies (Mackenzie et al. 1996; Miller et al. 2007; Miller et al. 2006), less is known about their specific functions. Two-hybrid and *in vitro* analyses demonstrated that NS4B interacted with NS3, displacing it from ssRNA and enhancing unwinding activity and replicase processivity on dsRNA (Umareddy et al. 2006). Furthermore, a genetic interaction between flavivirus NS1 and NS4A was shown to be critical for replicase activity (Lindenbach and Rice 1999). Lastly, NS2A has been demonstrated to play a role in WNV RNA synthesis (Mackenzie et al. 1998), suggesting that it may be involved in DENV RNA synthesis as well. Further analyses will clarify the precise roles of other DENV NS proteins in formation of the DENV RC and the mechanism of RNA synthesis.

3.4 Host Cell Factors Involved in Regulation of DENV Replication

Although host cell proteins have been shown to enhance DENV RNA synthesis in an *in vitro* extract (Raviprakash et al. 1998), at the time this review was written, no specific host-derived factors had been demonstrated to directly impact DENV RNA replication. As noted previously, several mammalian proteins including eIF1A, La, PTB, YB1, hnRNP A1, hnRNP A2/B1 and hnRNP Q bind to the 3′UTR of the DENV genome (Blackwell and Brinton 1997; De Nova-Ocampo et al. 2002; Paranjape and Harris 2007) (Fig. 2). Of these, La was shown to interact with the DENV 5′UTR, 3′UTR and the negative-strand 3′UTR, suggesting that it might play a role in positive- and negative-strand RNA synthesis (De Nova-Ocampo et al. 2002; Garcia-Montalvo et al. 2004; Yocupicio-Monroy et al. 2007; Yocupicio-Monroy et al. 2003). Biochemical evidence indicates that La interacts with NS3 and NS5 (Garcia-Montalvo et al. 2004) and undergoes nuclear localization upon DENV infection in C6/36 cells (Yocupicio-Monroy et al. 2007). However, although La has been shown to regulate HCV RNA synthesis (Domitrovich et al. 2005), experiments using a dominant-negative La protein revealed no impact of La on DENV replication (Walker, Paranjape, and Harris, unpublished). eIF1A has been shown to bind to the 3'SL of many flaviviruses, including DENV and WNV and to enhance negative-strand WNV RNA synthesis (Blackwell and Brinton 1997; Davis et al. 2007), suggesting a similar role for this protein in DENV replication. Indirect evidence indicating that RNA elements within the DENV 3'SL that mediate eIF1A binding are important for viral viability (Yu et al. 2008; Zeng et al. 1998) supports the idea that eIF1A is involved in DENV replication. Clearly, it will be important to extend our understanding of the role of host proteins in the regulation of viral RNA synthesis.

4 Methodologic Advances

4.1 *Proteomic and Biochemical Analysis of DENV-Host Interaction*

The development of *in vitro* assays has enabled reconstitution of DENV translation and negative-strand replication and the dissection of molecular mechanisms of DENV infection. Extension of these assays to determine the requirements for RF to RI conversion and plus-strand amplification will be critical. In addition, *in vitro* assays will continue to facilitate elucidation of the factors and parameters involved in the switch from translation to replication and between different stages of the viral lifecycle.

While viral sequence requirements for DENV translation and replication have been well-defined, there is much to learn about the host cell proteins involved in these processes. Over the past 5 years, progress in genomic and proteomic analysis has enabled insight into the genes and proteins that are up- or down-regulated during DENV infection (Fink et al. 2007; Hibberd et al. 2006; Pattanakitsakul et al. 2007). To obtain information about the proteins involved in regulation of translation and replication, basic biochemical and proteomic approaches must be combined. Using RNA-affinity chromatography and cross-linking strategies, proteins that associate with DENV regulatory sequences in reconstituted *in vitro* systems and cells can be isolated. Recent developments in proteomic approaches such as MUD-PIT mass spectrometry now allow efficient identification of these DENV binding factors. An important focus of these technologies in the future will be to delineate regulatory factors that modulate DENV translation and replication in mammalian and mosquito cells.

4.2 *Microscopic and Cell Biologic Methods to Discern Viral Dynamics*

Although biochemical methods are indispensable for determining protein-RNA interactions, understanding the molecular processes of DENV infection necessitates a comprehensive picture of the interactions within a cell. To this end, advances in immunoflourescence and electron microscopy and advances in techniques to label intracellular viral RNA (Dahm et al. 2008) will continue to extend our understanding of the subcellular localization of DENV translation and replication and our insight into the dynamics of replication foci. Moreover, techniques such as atomic force microscopy and fluorescence resonance energy transfer (FRET) that enable analysis of protein-RNA interactions at the molecular level can provide detailed information on molecular interactions that occur during viral infection.

4.3 Reverse Genetics

While identification of host DENV-interacting factors can be streamlined using a proteomic approach, determination of the role that these factors play in DENV translation, replication or other stages of the viral life cycle is often hampered by the lack of appropriate tools. Fortunately, the current availability of genetically deficient mice and siRNA libraries to thousands of genes significantly facilitates analysis of protein function. Similarly, completion of the *Ae. aegypti* genome sequence (Nene et al. 2007) and the continued development of mosquito stocks harboring gene deletions, disruptions and mutations will enable us to expand our understanding of the molecular details of DENV replication in the mosquito vector. In the meantime, similarities between mosquito and *Drosophila melanogaster* may prove useful in identifying and characterizing the role of host factors in regulation of viral replication. Finally, reverse genetics techniques have been immensely useful for analysis of *cis*-acting viral sequences and structures, and the increased availability of DENV replicons, replicon viral particles (RVPs) and infectious clones will further facilitate mechanistic analyses of translation and replication in the future.

5 DENV Translation and RNA Synthesis in the Context of the Viral Lifecycle: Perspectives

5.1 Regulatory Sequences within Coding Regions

Despite a detailed understanding of the role of the sequences and structures in the DENV UTRs, the contribution of coding region structures and sequences to viral replication has been under-explored. The recent finding that the conserved cHP structure within the capsid coding region regulates both DENV translation start site selection and RNA synthesis ((Clyde and Harris 2006), Clyde et al. 2008) indicates that coding region elements play important roles in the DENV lifecycle. Additional coding region regulatory elements have been identified whose disruption interferes with viral replication; delineation of which stages of the lifecycle are specifically affected is currently underway (Carmona, Clyde and Harris, unpublished).

5.2   vRNP Structures and Functions

An emerging trend has appeared with the recent application of biochemical and proteomic approaches to identify host proteins that interact with viral genomes. Specifically, it is becoming increasingly clear that hnRNPs bind with high specificity and affinity to the UTRs of DENV and other viruses (Kim et al. 2007; Paranjape and Harris 2007). Formation of viral ribonucleo-protein complexes (vRNPs) may prevent rapid degradation that would ensue if a naked viral genome was released into the cell cytoplasm. However, given the pleiotropic functions of the hnRNPs identified, we speculate that, in addition to stabilizing the RNA, these factors likely regulate various steps of the viral lifecycle. The formation of vRNPs may establish the genomic architecture and facilitate various processes of the viral lifecycle in a manner similar to the regulation of DNA transcription and replication by modification of chromatin components. Molecular genetic analysis of the roles of specific hnRNPs in regulation of the DENV lifecycle should enable a better understanding not only of regulation of DENV replication but also of the complex interaction between DENV and its host cell components.

5.3   Modulation of Host Cell Stress Responses by DENV

Stress granules are characterized by accumulation of proteins including TIA-1, TIAR, G3BP and YB-1 into discrete foci in the perinuclear space during periods of cellular stress when cap-dependent translation initiation is inhibited. Stress granules are sites of mRNA triage, while processing (P) bodies are sites of mRNA decay and turnover. Data from our laboratory and others have indicated that DENV infection represses the formation of stress granules and P bodies in certain cell types ((Emara and Brinton 2007); Paranjape and Harris, unpublished). While the precise mechanism for inhibition of stress granules and P bodies remains unclear, the inability of DENV-infected cells to form stress granules provides an explanation for the observation that flaviviruses, despite inducing anti-viral responses, do not repress host cell translation.

5.4   Switch Between Replication and Translation

Positive-sense RNA viral genomes must undergo numerous molecular processes on a single RNA molecule before viral packaging and export. Thus, input-strand translation and minus-strand replication must occur in opposite directions on each

DENV RNA template. Studies of poliovirus indicate that a switch is established when the viral protein, 3CD, binds the 5′UTR and represses translation to enable RNA synthesis (Gamarnik and Andino 1998). In the case of DENV, results indicating that NS5 binds to the DENV 5′SLA in a region that is likely to inhibit cap-dependent initiation suggest that NS5, possibly regulated by its phosphorylation state, may facilitate a switch between DENV translation and replication (Filomatori et al. 2006). Transitions of DENV genome architecture and circularization induced by protein binding and annealing of complementary conserved sequences (Fig. 3) are likely to play an important role in this regulation (Harris et al. 2006). Additionally, although studies of polio and Kunjin viruses suggest that only molecules that have successfully been replicated and translated are packaged (Khromykh et al. 2001b; Nugent et al. 1999), the requirements for DENV packaging have not been determined.

6 Conclusions

The establishment of DENV infection requires temporal, spatial and mechanistic coordination of the molecular processes of viral replication. Sequences and secondary structures within the DENV genome dynamically interact with viral and host proteins to regulate translation and replication. While we have detailed knowledge of the *cis* and *trans* elements required for DENV protein and RNA synthesis, many questions about the coordination of these processes remain to be answered. Continuing development of molecular, biochemical and cell biological tools will facilitate elucidation of the precise mechanisms of first- and second-round translation and of minus- and plus-strand replication while advances in proteomics and genomics are likely to provide information about the role of host factors in these processes. The DENV field is at an exciting point where new tools and approaches are advancing our understanding of the intricacies of DENV translation and replication.

References

Ackermann M, Padmanabhan R (2001) J Biol Chem 276:39926–39937
Alvarez DE, De Lella Ezcurra AL, Fucito S, Gamarnik AV (2005a) Virology 339:200–212
Alvarez DE, Filomatori CV, Gamarnik AV (2008) Functional analysis of dengue virus cyclization sequences located at the 5′ and 3′UTRs. Virology 375:223–235
Alvarez DE, Lodeiro MF, Filomatori CV, Fucito S, Mondotte JA, Gamarnik AV (2006) Novartis Found Symp 277:120–132 discussion 132–135, 251–253
Alvarez DE, Lodeiro MF, Luduena SJ, Pietrasanta LI, Gamarnik AV (2005b) J Virol 79:6631–6643
Bartholomeusz AI, Wright PJ (1993) Arch Virol 128:111–121
Blackwell JL, Brinton MA (1997) J Virol 71:6433–6444

Blaney JE Jr, Hanson CT, Firestone CY, Hanley KA, Murphy BR, Whitehead SS (2004) Am J Trop Med Hyg 71:811–821

Blaney JE Jr, Sathe NS, Goddard L, Hanson CT, Romero TA, Hanley KA, Murphy BR, Whitehead SS (2008) Vaccine 26:817–828

Brinton MA, Dispoto JH (1988) Virology 162:290–299

Butrapet S, Huang CY, Pierro DJ, Bhamarapravati N, Gubler DJ, Kinney RM (2000) J Virol 74:3011–3019

Cahour A, Pletnev A, Vazielle-Falcoz M, Rosen L, Lai CJ (1995) Virology 207:68–76

Chiu WW, Kinney RM, Dreher TW (2005) J Virol 79:8303–8315

Cleaves GR, Dubin DT (1979) Virology 96:159–165

Cleaves GR, Ryan TE, Schlesinger RW (1981) Virology 111:73–83

Clyde K, Harris E (2006) J Virol 80:2170–2182

Clyde K, Barrera J, Harris E (2008) Virology 379:314–323

Clyde K, Kyle JL, Harris E (2006) J Virol 80:11418–11431

Cui T, Sugrue RJ, Xu Q, Lee AK, Chan YC, Fu J (1998) Virology 246:409–417

Dahm R, Zeitelhofer M, Gotze B, Kiebler MA, Macchi P (2008) Methods Cell Biol 85:293–327

Davis WG, Blackwell JL, Shi PY, Brinton MA (2007) J Virol 81:10172–10187

De Nova-Ocampo M, Villegas-Sepulveda N, del Angel RM (2002) Virology 295:337–347

Domitrovich AM, Diebel KW, Ali N, Sarker S, Siddiqui A (2005) Virology 335:72–86

Dong H, Ray D, Ren S, Zhang B, Puig-Basagoiti F, Takagi Y, Ho CK, Li H, Shi PY (2007) J Virol 81:4412–4421

Durbin AP, McArthur J, Marron JA, Blaney JE Jr, Thumar B, Wanionek K, Murphy BR, Whitehead SS (2006) Hum Vaccin 2:167–173

Durbin AP, Whitehead SS, McArthur J, Perreault JR, Blaney JE Jr, Thumar B, Murphy BR, Karron RA (2005) J Infect Dis 191:710–718

Edgil D, Diamond MS, Holden KL, Paranjape SM, Harris E (2003) Virology 317:275–290

Edgil D, Harris E (2006) Virus Res 119:43–51

Edgil D, Polacek C, Harris E (2006) J Virol 80:2976–2986

Egloff MP, Benarroch D, Selisko B, Romette JL, Canard B (2002) Embo J 21:2757–2768

Emara MM, Brinton MA (2007) Proc Natl Acad Sci USA 104:9041–9046

Filomatori CV, Lodeiro MF, Alvarez DE, Samsa MM, Pietrasanta L, Gamarnik AV (2006) Genes Dev 20:2238–2249

Fink J, Gu F, Ling L, Tolfvenstam T, Olfat F, Chin KC, Aw P, George J, Kuznetsov VA, Schreiber M, Vasudevan SG, Hibberd ML (2007) PLoS Negl Trop Dis 1:e86

Gamarnik AV, Andino R (1998) Genes Dev 12:2293–2304

Garcia-Montalvo BM, Medina F, del Angel RM (2004) Virus Res 102:141–150

Hahn CS, Hahn YS, Rice CM, Lee E, Dalgarno L, Strauss EG, Strauss JH (1987) J Mol Biol 198:33–41

Harris E, Holden KL, Edgil D, Polacek C, Clyde K (2006) Novartis Found Symp 277:23–39 discussion 40, 71–73, 251–253

Hibberd ML, Ling L, Tolfvenstam T, Mitchell W, Wong C, Kuznetsov VA, George J, Ong SH, Ruan Y, Wei CL, Gu F, Fink J, Yip A, Liu W, Schreiber M, Vasudevan SG (2006) Novartis Found Symp 277:206–214 discussion 214–217, 251–253

Holden KL, Harris E (2004) Virology 329:119–133

Holden KL, Stein DA, Pierson TC, Ahmed AA, Clyde K, Iversen PL, Harris E (2006) Virology 344:439–452

Kapoor M, Zhang L, Mohan PM, Padmanabhan R (1995) Gene 162:175–180

Keller TH, Chen YL, Knox JE, Lim SP, Ma NL, Patel SJ, Sampath A, Wang QY, Yin Z, Vasudevan SG (2006) Novartis Found Symp 277:102–114 discussion 114–9, 251–253

Khromykh AA, Kondratieva N, Sgro JY, Palmenberg A, Westaway EG (2003) J Virol 77:10623–10629

Khromykh AA, Meka H, Guyatt KJ, Westaway EG (2001a) J Virol 75:6719–6728

Khromykh AA, Sedlak PL, Westaway EG (2000) J Virol 74:3253–3263

Khromykh AA, Varnavski AN, Sedlak PL, Westaway EG (2001b) J Virol 75:4633–4640

Khromykh AA, Westaway EG (1997) J Virol 71:1497–1505

Kim CS, Seol SK, Song OK, Park JH, Jang SK (2007) J Virol 81:3852–3865

Kinney RM, Huang CY, Rose BC, Kroeker AD, Dreher TW, Iversen PL, Stein DA (2005) J Virol 79:5116–5128

Lindenbach BD, Rice CM (1999) J Virol 73:4611–4621

Lindenbach BD, Rice CM (2003) Adv Virus Res 59:23–61

Lo MK, Tilgner M, Bernard KA, Shi PY (2003) J Virol 77:10004–10014

Mackenzie JM, Jones MK, Young PR (1996) Virology 220:232–240

Mackenzie JM, Khromykh AA, Jones MK, Westaway EG (1998) Virology 245:203–215

Markoff L (2003) Adv Virus Res 59:177–228

Matusan AE, Pryor MJ, Davidson AD, Wright PJ (2001) J Virol 75:9633–9643

Miller S, Kastner S, Krijnse-Locker J, Buhler S, Bartenschlager R (2007) J Biol Chem 282:8873–8882

Miller S, Sparacio S, Bartenschlager R (2006) J Biol Chem 281:8854–8863

Mohan PM, Padmanabhan R (1991) Gene 108:185–191

Nene V, Wortman JR, Lawson D, Haas B, Kodira C, Tu ZJ, Loftus B, Xi Z, Megy K, Grabherr M, Ren Q, Zdobnov EM, Lobo NF, Campbell KS, Brown SE, Bonaldo MF, Zhu J, Sinkins SP, Hogenkamp DG, Amedeo P, Arensburger P, Atkinson PW, Bidwell S, Biedler J, Birney E, Bruggner RV, Costas J, Coy MR, Crabtree J, Crawford M, Debruyn B, Decaprio D, Eiglmeier K, Eisenstadt E, El-Dorry H, Gelbart WM, Gomes SL, Hammond M, Hannick LI, Hogan JR, Holmes MH, Jaffe D, Johnston JS, Kennedy RC, Koo H, Kravitz S, Kriventseva EV, Kulp D, Labutti K, Lee E, Li S, Lovin DD, Mao C, Mauceli E, Menck CF, Miller JR, Montgomery P, Mori A, Nascimento AL, Naveira HF, Nusbaum C, O'Leary S, Orvis J, Pertea M, Quesneville H, Reidenbach KR, Rogers YH, Roth CW, Schneider JR, Schatz M, Shumway M, Stanke M, Stinson EO, Tubio JM, Vanzee JP, Verjovski-Almeida S, Werner D, White O, Wyder S, Zeng Q, Zhao Q, Zhao Y, Hill CA, Raikhel AS, Soares MB, Knudson DL, Lee NH, Galagan J, Salzberg SL, Paulsen IT, Dimopoulos G, Collins FH, Birren B, Fraser-Liggett CM, Severson DW (2007) Genome sequence of Aedes aegypti, a major arbovirus vector. Science 316:1718–1723

Nomaguchi M, Ackermann M, Yon C, You S, Padmanabhan R (2003) J Virol 77:8831–8842

Nugent CI, Johnson KL, Sarnow P, Kirkegaard K (1999) J Virol 73:427–435

Paranjape SM, Harris E (2007) J Biol Chem 282:30497–30508

Patkar CG, Kuhn RJ (2006) Novartis Found Symp 277:41–52 discussion 52–56, 71–73, 251–253

Pattanakitsakul SN, Rungrojcharoenkit K, Kanlaya R, Sinchaikul S, Noisakran S, Chen ST, Malasit P, Thongboonkerd V (2007) J Proteome Res 6:4592–4600

Polacek C, Foley J, Harris E (2009a) J Virol 83:1161–1166

Polacek C, Friebe P, Harris E (2009b) J Gen Virol 90:687–692

Raviprakash K, Sinha M, Hayes CG, Porter KR (1998) Am J Trop Med Hyg 58:90–95

Sampath A, Xu T, Chao A, Luo D, Lescar J, Vasudevan SG (2006) J Virol 80:6686–6690

Selisko B, Dutartre H, Guillemot JC, Debarnot C, Benarroch D, Khromykh A, Despres P, Egloff MP, Canard B (2006) Virology 351:145–158

Shurtleff AC, Beasley DW, Chen JJ, Ni H, Suderman MT, Wang H, Xu R, Wang E, Weaver SC, Watts DM, Russell KL, Barrett AD (2001) Virology 281:75–87

Tajima S, Nukui Y, Takasaki T, Kurane I (2007) J Gen Virol 88:2214–2222

Tan BH, Fu J, Sugrue RJ, Yap EH, Chan YC, Tan YH (1996) Virology 216:317–325

Tilgner M, Deas TS, Shi PY (2005) Virology 331:375–386

Umareddy I, Chao A, Sampath A, Gu F, Vasudevan SG (2006) J Gen Virol 87:2605–2614

Westaway EG, Khromykh AA, Mackenzie JM (1999) Virology 258:108–117

Westaway EG, Mackenzie JM, Kenney MT, Jones MK, Khromykh AA (1997) J Virol 71:6650–6661

Westaway EG, Mackenzie JM, Khromykh AA (2002) Curr Top Microbiol Immunol 267:323–351

Whitehead SS, Falgout B, Hanley KA, Blaney JE Jr, Markoff L, Murphy BR (2003) A live, attenuated dengue virus type 1 vaccine candidate with a 30-nucleotide deletion in the 3' untranslated region is highly attenuated and immunogenic in monkeys. J Virol 77:1653–1657

Xu T, Sampath A, Chao A, Wen D, Nanao M, Luo D, Chene P, Vasudevan SG, Lescar J (2006) Novartis Found Symp 277:87–97 discussion 97–101, 251–253

Yocupicio-Monroy M, Padmanabhan R, Medina F, del Angel RM (2007) Virology 357:29–40

Yocupicio-Monroy RM, Medina F, Reyes-del Valle J, del Angel RM (2003) J Virol 77:3067–3076

You S, Falgout B, Markoff L, Padmanabhan R (2001) J Biol Chem 276:15581–15591

Yu L, Nomaguchi M, Padmanabhan R, Markoff L (2008) Specific requirements for elements of the 5' and 3' terminal regions in flavivirus RNA synthesis and viral replication. Virology 374:170–185

Zeng L, Falgout B, Markoff L (1998) Identification of specific nucleotide sequences within the conserved 39-SL in the dengue type 2 virus genome required for replication. J Virol 72:7510–7522

Zhang B, Dong H, Stein DA, Iversen PL, Shi PY (2008) West Nile virus genome cyclization and RNA replication require two pairs of long-distance RNA interactions. Virology 373:1–13

Subversion of Interferon by Dengue Virus

Jorge L. Muñoz-Jordán

Contents

Abstract Dengue virus is sensed in mammalian cells by Toll-like receptors and DExD/H box RNA helicases, triggering a Type 1 interferon response. Interferon acts upon infected and noninfected cells by stimulating the JAK/STAT signaling pathway resulting in the activation of interferon stimulated genes that lead cells toward the establishment of an antiviral response. The recognition of the importance of this rapid protective response should come with the realization that dengue virus would circumvent the interferon response to propagate in the host. There is recent, mounting evidence for mechanisms encoded by the dengue virus that weaken interferon signaling. Nonstructural proteins expressed separately or in replicon vectors block phosphorylation and down-regulate expression of major components of the JAK/STAT pathway, causing reduced activation of gene expression in response to IFNα/β interferon. As our understanding of viral-host interaction

J.L. Muñoz-Jordán

Centers for Disease Control and Prevention, Division of Vector Borne Infectious Diseases, Dengue Branch, 1324 Calle Cañada, San Juan, PR 00920, USA
e-mail: ckq2@cdc.gov

A.L. Rothman (ed.), *Dengue Virus*, Current Topics in Microbiology and Immunology 338,   35
DOI 10.1007/978-3-642-02215-9_3, © Springer-Verlag Berlin Heidelberg 2010

increases, opportunities for improved biological models and therapeutics discovery arise.

1 Introduction

Despite the undoubted importance of the dengue virus (DENV) as a human pathogen, no immunization or therapeutic interventions are yet available to control this disease. The progress in developing vaccines and antivirals has been hindered by the lack of biological models that mimic disease progression. Biological assays are the only options for the study of host/virus interactions; but the validity of experimental data is often fraught with the difficulties inherent in clinical studies. Type 1 interferon (IFNα/β) antagonism has emerged in recent years as a rapidly growing area of research because of the significant role that these cytokines play in establishing protection against pathogens (Garcia-Sastre and Biron 2006; Pitha and Kunzi 2007). IFNα/β products are made and secreted by mammalian cells within just hours, sometimes minutes, of the cells' becoming infected by viruses, producing in the invaded and neighboring cells a refractory state that protects the host against viral intruders. The differential specificity of multiple sensing mechanisms for viral components ensures that mammalian cells are prepared to recognize viruses that greatly differ in their structural and functional features. This diversity is further enhanced by the assortment of IFN pathways activated during viral infection (Onomoto et al. 2007; Severa and Fitzgerald 2007). DENV activates a complex and flexible antiviral response; but this response is counteracted by this virus as it remains one step ahead of the host's ability to wipe it out. Amidst this elaborate backdrop, a picture is emerging with increasing clarity of the molecular mechanisms underlying subversion of the IFNα/β response by DENV. The findings are likely to impact our understanding of dengue pathogenesis and help advance the development of antiviral approaches.

2 Sensing Dengue

Viruses are detected by at least two major classes of pathogen recognition receptors: the extracellular/endosomal Toll-like receptors (TLRs) (Akira and Takeda 2004; Bowie and Haga 2005) and the cytoplasmic receptor family of DExD/H box RNA helicases (e.g., RIG-I: retinoic acid inducible gene I and MDA5: melanoma differentiation-associated gene-5) (Meylan and Tschopp 2006). DENV is sensed by both of these mechanisms. The endosomal side of TLR3, 7, 8 and 9 is involved in detecting viral nucleic acids. The intracellular regions of TLRs contain a structural motif common to all TLRs and IL-1 receptors known as TLRs/IL-1 resistance (TIR) domain, which recruits adaptor molecules, leading to activation of transcription factors that induce IFNα/β expression. TLR7 is found mainly in endosomal compartments of dendritic cells (DC), which release high levels of IFNα/β to the serum (Diebold et al., 2004; Lund et al. 2004). It is clear that DENV activates DC and induces production of IFNα/β, TNFα and a strong pro-inflammatory response (Libraty et al. 2001; Wang et al. 2006). Circulating myeloid DC and plasmacytoid DC (pDC) decrease in number and frequency during acute dengue illnesses in pediatric patients. This decrease correlates with the increased severity of the disease and, remarkably, an early decrease in circulating DC levels was found in patients who subsequently developed dengue hemorrhagic fever (DHF) (Pichyangkul et al. 2003). Recognition of DENV RNA by TLR7 is linked to viral fusion and uncoating processes that occur in endosomal compartments during early infection of pDCs, leading to a robust IFNα/β production (Fig. 1a). The signaling process is enhanced

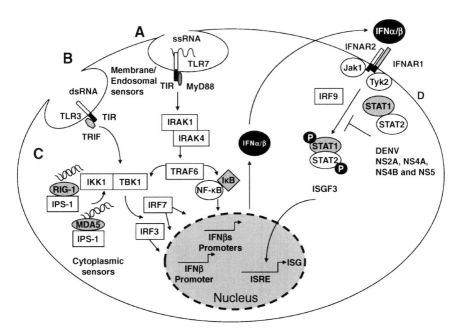

Fig. 1 Sensing Dengue Virus Infection by the Host Cell. (**a**) Recognition of DENV RNA by TLR7 occurs in endosomal compartments during early infection of dendritic cells, leading to IFNα/β production. Induction of IFNα/β through TLR7/8/9 is mediated by the adaptor molecule myeloid differentiation primary response protein 88 (MyD88), resulting in downstream activation of IRF7, IKKα/β/γ and MAPK cascades, leading to activation of NFκB and AP-1. IRF7 activates IFNα and IFNβ expression. (**b**)TLR3, a broad sensor of dsRNA intermediates during viral infections, is upregulated in human umbilical endothelial cells (HUVEC) by DENV infection. IRF3 leads to activation of IFNβ. (**c**) Cytoplasmic sensing of DENV occurs through RIG-I and MDA5, resulting in activation of a macromolecular signaling complex that stimulates IRF3; in turn this activation induces the IFNβ promoter. (**d**) Secreted IFNα/β activate the IFNα/β receptor and the JAK/STAT pathway, leading to phosphorylation and dimerization of STAT1/2 and formation of the macro-molecular factor ISGF3, which translocates to the nucleus and activates ISREs. The nonstructural proteins NS4B, NS5 and to a lesser extent NS2A and NS4A, impair parts of the JAK/STAT pathway and reduce activation of ISREs

by the higher order structure of viral RNA, indicating the importance of viral structural motifs in recognition by TLRs (Wang et al. 2006). The role of TLR8/9, if any, has not yet been defined for dengue. However, these receptors are involved in activation of DCs by yellow fever virus to produce IFNα and proinflammatory cytokine responses (Querec et al. 2006). Regardless, induction of IFNα/β through TLR7/8/9 is generally mediated by the adaptor molecule myeloid differentiation primary response protein 88 (MyD88), resulting in downstream activation of IRF7, IKKα/β/γ and MAPK cascades, leading to activation of NFκB and AP-1. Evidence for activation of this pathway by RNA viruses and in particular DENV, has been more recently recognized (Chang et al. 2006; Loo et al. 2008). IRF7 activates IFNα and IFNβ expression, whereas IRF3 leads to activation of IFNβ. Then, pDCs, natural targets of DENV infection constitutively expressing TLR7 and IRF7, rapidly lead to a systemic IFNα/β response with dengue infection. TLR3 is a broad sensor of dsRNA, a by-product of ssRNA viral replication and transcription and is upregulated in human umbilical endothelial cells (HUVEC) by DENV infection (Warke et al. 2003) (Fig. 1b).

Cytoplasmic sensing of DENV has also been documented. RIG-I and MDA5 signaling occurs through interactions between their homotypic caspase activation and recruitment domain (CARD) and the IPS-1/MAVS/CARDIF adaptor protein in association with the outer mitochondrial membrane, resulting in activation of a macromolecular signaling complex that stimulates IRF3; in turn this activation induces the IFNβ promoter. RIG-I and MDA5 share similar signaling features and structural homology but the two helicases discriminate among different ligands. While RIG-I recognizes RNA secondary structures and ssRNA 5′-triphosphate ends, MDA5 recognizes annealing inosine- and cytosine-containing RNA strands. Recent experiments using single and double RIG-I/MDA5 knockout mouse fibroblast cell lines show that DENV triggers both of these responses (Chang et al. 2006; Loo et al. 2008) (Fig. 1c).

3 Induction of IFNα/β by DENV

DENV induces a strong IFNα/β response in natural infections. This fact has been well established by a study of healthy and dengue pediatric patients. Moreover, IFNα/β is present at higher levels and for a longer period of time after defervescence in some pediatric DHF patients than in dengue fever patients (Kurane et al. 1993). The fundamental role of IFNα/β in protecting against DENV infections had been exemplified by the demonstration that directed impairment of both IFNα/β signaling in mice (IFNAR$^{-/-}$ IFNGR$^{-/-}$ mice) results in high lethality and consistent viremia in serum and tissue (Johnson and Roehrig 1999). Genomics technologies such as microarrays have revealed patterns of IFN signaling in cells infected with a variety of pathogens, allowing researchers to explore the complexity of the IFNα/β response on a large-scale (Jenner and Young 2005). Due to the lack of animal models by which to study dengue disease and the complex virus/host interactions leading to IFN modulation, investigation has turned toward the study of these large gene expression profiles and the relevance of these findings in dengue natural infections or vaccine experimentation. These studies revealed upregulation of the key mediators of inflammatory cytokine responses, antiviral response through IFNα/β activation, NF-κB-mediated cytokine/chemokine responses and the ubiquitin proteosome pathway (Fink et al. 2007; Sariol et al. 2007; Warke et al. 2003). TNF-related apoptosis-inducing ligand (TRAIL) is upregulated in HUVEC, DC and monocytes infected by DENV and it may play a role controlling DENV infection (Warke et al. 2008). Since TRAIL expression is induced by IFNα/β signaling, its role as part of the innate antiviral response should be further analyzed. The chemokines IP-10 and I-TAC (both ligands of the CXCR3 receptor) are also highly upregulated genes in microarrays of patients with severe dengue at the early onset of fever (Fink et al. 2007). Viperin is highly upregulated during DENV infections and induced by IFNα/β and has been picked up by human and rhesus microarrays (Fink et al. 2007; Sariol et al. 2007). Its over-expression apparently reduces viral replication; but it would be important to discern this inhibition from a direct effect by the IFN present in the system. Considering that endothelium permeability is a major factor contributing to DHF and that DC are activated by DENV, the conserved features of IFNα/β and proinflammatory responses observed in these microarrays underscore the importance of innate immune responses during DENV infection; however, interspecific differences may indicate unrecognized aspects of DENV pathogenesis (Sariol et al. 2007).

4 Inhibition of IFN α/β signaling by DENV

Secreted IFNα/β binds to the IFNα/β receptor (IFNAR) found on the surface of infected and neighboring cells, resulting in activation of the Janus kinase (JAK)/ signal transducer and activator of transcription (STAT) pathway and transcription of numerous genes from promoters containing IFN-stimulated regulatory elements (ISRE) (Takaoka and Yanai 2006). This process results in the induction of expression of hundreds of genes and the establishment of an antiviral state (Der et al. 1998). Experimentally, DENV induces IFNα/β and is susceptible to the action of IFN. Pretreatment of human hepatoma cells with IFNβ inhibits DENV replication. The inhibitory effect of IFNβ is retained even when DENV RNA is transfected directly into cells, indicating that IFNβ inhibits postentry steps of viral replication. This inhibition appears to be independent of PKR and RNase L, because genetically deficient PKR- RNase L- cells that were infected by DENV retained sensitivity to IFNβ (Diamond and Harris 2001).

Abrogation of the IFNα/β receptor and JAK/STAT pathway results in increased viremia in experimental models suggesting the importance of the early immune response in inhibiting DENV replication (Johnson and Roehrig 1999; Shresta et al. 2006; Shresta et al. 2005). In addition, clinical studies demonstrated that IFNα/β peaks very early in natural DENV infections, with IFN levels decreasing with disease progression (Libraty et al. 2002) and DC activation occurring prior to detection of DENV (Libraty et al. 2001; Pichyangkul et al. 2003). These observations could be suggesting that DENV limits the effects of IFN and propagates in the presence of a full-bodied antiviral response. In addition, abrogation of IFNα/β receptor and JAK/STAT pathway results in increased viremia in experimental models (Johnson and Roehrig 1999; Shresta et al. 2006; Shresta et al. 2005). In addition, infection by DENV just prior to IFNβ treatment of cells in vitro blocks the inhibitory effect of IFNβ, confirming that the action of IFNβ upon infected cells is hindered by the virus (Diamond and Harris 2001). Such a hindering of IFNβ could be the explanation for the enhanced levels of dengue viremia seen to accompany the aforementioned impairment of IFN signaling. In vitro infections of LLCMK2 cells with DENV showed that STAT1 was not phosphorylated or localized in the nucleus of infected cells. Of 10 DENV-encoded proteins expressed separately in human alveolar basal epithelial cells (A549), NS2A, NS4A and NS4B enhanced replication of IFN-sensitive viruses in the presence of exogenous IFNβ (Munoz-Jordan et al. 2003). Through the appropriate use of in vitro systems, it is possible to discern different parts of the IFNα/β pathways as means of narrowing down the interactions with the virus. Induction of IFNα/β in response to viral infection can be studied in 293T cells separate from IFN stimulation, whereas Vero cells, which do not produce IFNβ, are used for the study of JAK/STAT signaling if physiological concentrations of IFNβ are present. These three DENV proteins inhibited ISRE promoters placed in reporter constructs following direct stimulation with IFNβ. These experiments have provided a quantitative measure of the inhibitory effects of these proteins against IFNα/β stimulation, with NS4B representing the strongest IFN antagonist, capable of inhibiting STAT1 phosphorylation and nuclear translocation upon stimulation with IFNβ (Munoz-Jordan et al. 2003). Coexpression of NS4A and NS4B resulted in enhanced inhibition of the ISRE promoter activation in response to IFNβ and a full IFN antagonistic effect was obtained if NS2A, NS4A and NS4B were expressed together (Munoz-Jordan et al. 2005; Munoz-Jordan et al. 2003). It is then possible that while these three proteins can individually produce partial inhibition of IFNα/β, they may act synergistically during viral infection upon components of the JAK/STAT pathway to completely subvert the IFNα/β response (Fig. 1d). None of the 10 DENV proteins appeared to inhibit IFN induction separately; however, it is possible that other DENV proteins different from NS2A, NS4A and NS4B participate in inhibition of IFN expression or IFN signaling through interactions among themselves. Because DENV-infected cells exhibit upregulation of IRF3 (Warke et al. 2003), direct stimulation of Interferon Sensitive Genes (ISGs) by ISRE may be possible and other mechanisms may have been developed by DENV to avert this pathway that are yet to be identified.

The role of nonstructural proteins of DENV in inhibiting JAK/STAT signaling was further documented with the use of dengue replicon-containing cells and DENV-infected cells, where IFN-sensitive virus was resistant to the effect of IFNα because of reduced levels of STAT2 (Jones et al. 2005). The virus polymerase NS5, binds to STAT2 and causes its reduced expression. This process requires ubiquitination and proteasome activity, strongly suggesting active degradation.

Binding of NS5 to STAT2 is dependent on the expression of NS5 in the context of a polyprotein that undergoes proteolytic processing for NS5 maturation. Thus, NS5 is able to target STAT2 for degradation only if it expressed as a precursor(Ashour et al. 2009). The observation that DENV blocked JAK/STAT signaling and probably not IFNα/β activation, is further substantiated by the demonstration that activation of ISRE promoters insensitive to IRFs was disrupted by DENV NS4B (Munoz-Jordan et al. 2005). The exact mechanism for JAK/STAT suppression by DENV has not yet been elucidated; but the reduced levels of STAT2 and the lack of STAT1 phosphorylation may be related to down-regulation of Tyk2 (Ho et al. 2005; Lin et al. 2004), with levels of IFNAR remaining unaltered (Ho et al. 2005; Jones et al. 2005). These observations potentially place these interactions in upstream sections of the JAK/STAT signaling pathway.

Folding of the DENV polyprotein in association with the ER membrane and timely posttranslational modifications are required for the completion of DENV life cycle. The NS4A/B region is cleaved by the host signal peptidase only after cleavage by the viral peptidase (NS2B/3) has occurred (Lin et al. 1993). IFN antagonistic function of the nonstructural region seems to be synchronized with this cleavage process, because NS4A/B fusion protein did not cause inhibition of IFNα/β until this product was cleaved by the viral protease NS2B/3. Also, deletion of NS4B signal peptide (2K segment) and non replacement by another signal peptide, showed that correct targeting of this protein to the ER membrane is required for this function. Additionally, a deletion analysis of NS4B showed that amino acids between residues 77 and 125 were required for IFNα/β antagonism (Munoz-Jordan et al. 2003). These residues are predicted to be located on the cytoplasmic surface between the first and second transmembrane regions of NS4B (Lundin et al. 2003; Qu et al. 2001).

The function of NS4B and other flavivirus nonstructural proteins in suppressing IFN JAK/STAT signaling has been documented in other flavivirueses in keeping with observations that these viruses have similar mechanisms to block this pathway (Guo et al. 2005; Keller et al. 2006; Liu et al. 2005; Munoz-Jordan et al. 2005). NS4B function does not seem to be conserved in tick-borne flaviviruses and Japanese encephalitis virus (JEV), where the viral replicase NS5 holds the ability to block JAK/STAT signaling (Best et al. 2005). The minimal requirement for the JAK/STAT inhibitory effect is retained within residues in two noncontiguous sequences of the RNA-dependent RNA polymerase region of JEV NS5, which seem to come together in the tridimensional structure of this protein. These features are partially conserved across flaviviruses, potentially contributing to the apparent differences observed in DENV (Park et al. 2007). Important interspecies and even intercellular differences need to be considered, as different players in the JAK/STAT pathway have been found to be antagonized by DENV (Jones et al. 2005; Munoz-Jordan et al. 2003).

It is possible that mechanisms may differ across flaviviruses or across DENV serotypes and even between clades of the same serotypes. Strain differences between pathogenic and nonpathogenic strains of DENV and WNV likely correlate with differences in their ability to block JAK/STAT signaling (Ho et al. 2005;

Keller et al. 2006). It might be that NS4B and other nonstructural proteins block IFN in ways reminiscent of paramyxovirus P gene products (named V, P and C), which block the JAK/STAT function in a variety of ways (Haller and Weber 2007).

It is tempting to speculate that the IFN-antagonistic function of DENV nonstructural proteins is most relevant in cells infected by this virus if the action of IFNα/β upon this cell is impaired. However, since RIG-I and MDA5 are IFN-inducible, important positive feedback signals may be inhibited by the nonstructural proteins of DENV as they block the JAK/STAT pathway, rendering cells less responsive to viral infection.

5 Conclusions and Perspectives

The last 10 years of research on the innate immunity against DENV has revealed the main features of IFNα/β modulation during DENV infection. Lack of animal models for the study of DENV pathogenesis force us to integrate in vitro, genomics and proteomics approaches and clinical/diagnostic studies to better understand the significance of these findings. DENV activates most of the cell's sensing mechanisms and triggers a robust IFNα/β response. But the antiviral effects of IFNα/β upon the infected cell are blocked by virus-encoded nonstructural proteins, causing down-regulation of the JAK/STAT pathway and impaired signaling of ISREs. The IFN response is complex and other interactions between DENV and the host antiviral response are likely to be unveiled in the near future. Substantial effort is being invested in understanding the precise molecular interactions between NS4B and the JAK/STAT pathway and how these mechanisms could play a role in disease. Strain differences that could correlate with differences in IFNα/β activation or inhibition are likely subjects of research. The analysis of large gene expression profiles linked to clinical studies may lead to the identification of specific factors with the potential for therapeutic intervention and correlation of protection in vaccine studies. The biological and medical relevance of these findings needs to be investigated with appropriately designed clinical studies to identify therapeutic targets.

Acknowledgments Adolfo Garcia-Sastre, Wellington Sun, Carlos Sariol, Luis Martinez, Daniel Libraty and Irene Bosch provided insightful comments.

References

Akira S, Takeda K (2004) Toll-like receptor signaling. Nat Rev Immunol 4:499–511
Ashour J, Laurent-Rolle M, Shi PY, Garcia-Sastre A (2007) NS5 of dengue virus mediates STAT2 binding and degradation. J Virol 83:5408–5418

Best SM, Morris KL, Shannon JG, Robertson SJ, Mitzel DN, Park GS, Boer E, Wolfinbarger JB, Bloom ME (2005) Inhibition of interferon-stimulated JAK-STAT signaling by a tick-borne flavivirus and identification of NS5 as an interferon antagonist. J Virol 79:12828–12839

Bowie AG, Haga IR (2005) The role of Toll-like receptors in the host response to viruses. Mol Immunol 42:859–867

Chang TH, Liao CL, Lin YL (2006) Flavivirus induces interferon-beta gene expression through a pathway involving RIG-I-dependent IRF-3 and PI3K-dependent NF-kappaB activation. Microbes Infect 8:157–171

Der SD, Zhou A, Williams BR, Silverman RH (1998) Identification of genes differentially regulated by interferon alpha, beta, or gamma using oligonucleotide arrays. Proc Natl Acad Sci USA 95:15623–15628

Diamond MS, Harris E (2001) Interferon inhibits dengue virus infection by preventing translation of viral RNA through a PKR-independent mechanism. Virology 289:297–311

Diebold SS, Kaisho T, Hemmi H, Akira S, ReiseSousa C (2004) Innate antiviral responses by means of TLR7-mediated recognition of single-stranded RNA. Science 303:1529–1531

Fink J, Gu F, Ling L, Tolfvenstam T, Olfat F, Chin KC, Aw P, George J, Kuznetsov VA, Schreiber M et al (2007) Host gene expression profiling of dengue virus infection in cell lines and patients. PLoS Negl Trop Dis 1:e86

Garcia-Sastre A, Biron CA (2006) Type 1 interferons and the virus-host relationship: a lesson in detente. Science 312:879–882

Guo JT, Hayashi J, Seeger C (2005) West Nile virus inhibits the signal transduction pathway of alpha interferon. J Virol 79:1343–1350

Haller O, Weber F (2007) Pathogenic viruses: smart manipulators of the interferon system. Curr Top Microbiol Immunol 316:315–334

Ho LJ, Hung LF, Weng CY, Wu WL, Chou P, Lin YL, Chang DM, Tai TY, Lai JH (2005) Dengue virus type 2 antagonizes IFN-alpha but not IFN-gamma antiviral effect via down-regulating Tyk2-STAT signaling in the human dendritic cell. J Immunol 174:8163–8172

Jenner RG, Young RA (2005) Insights into host responses against pathogens from transcriptional profiling. Nat Rev Microbiol 3:281–294

Johnson AJ, Roehrig JT (1999) New mouse model for dengue virus vaccine testing. J Virol 73:783–786

Jones M, Davidson A, Hibbert L, Gruenwald P, Schlaak J, Ball S, Foster GR, Jacobs M (2005) Dengue virus inhibits alpha interferon signaling by reducing STAT2 expression. J Virol 79:5414–5420

Keller BC, Fredericksen BL, Samuel MA, Mock RE, Mason PW, Diamond MS, Gale M Jr (2006) Resistance to alpha/beta interferon is a determinant of West Nile virus replication fitness and virulence. J Virol 80:9424–9434

Kurane I, Dai LC, Livingston PG, Reed E, Ennis FA (1993) Definition of an HLA-DPw2-restricted epitope on NS3, recognized by a dengue virus serotype-cross-reactive human CD4+ CD8- cytotoxic T-cell clone. J Virol 67:6285–6288

Libraty DH, Pichyangkul S, Ajariyakhajorn C, Endy TP, Ennis FA (2001) Human dendritic cells are activated by dengue virus infection: enhancement by gamma interferon and implications for disease pathogenesis. J Virol 75:3501–3508

Libraty DH, Young PR, Pickering D, Endy TP, Kalayanarooj S, Green S, Vaughn DW, Nisalak A, Ennis FA, Rothman AL (2002) High circulating levels of the dengue virus nonstructural protein NS1 early in dengue illness correlate with the development of dengue hemorrhagic fever. J Infect Dis 186:1165–1168

Lin C, Amberg SM, Chambers TJ, Rice CM (1993) Cleavage at a novel site in the NS4A region by the yellow fever virus NS2B–3 proteinase is a prerequisite for processing at the downstream 4A/4B signalase site. J Virol 67:2327–2335

Lin RJ, Liao CL, Lin E, Lin YL (2004) Blocking of the alpha interferon-induced Jak-Stat signaling pathway by Japanese encephalitis virus infection. J Virol 78:9285–9294

Liu WJ, Wang XJ, Mokhonov VV, Shi PY, Randall R, Khromykh AA (2005) Inhibition of interferon signaling by the New York 99 strain and Kunjin subtype of West Nile virus involves blockage of STAT1 and STAT2 activation by nonstructural proteins. J Virol 79:1934–1942

Loo YM, Fornek J, Crochet N, Bajwa G, Perwitasari O, Martinez-Sobrido L, Akira S, Gill MA, Garcia-Sastre A, Katze MG, Gale M Jr (2008) Distinct RIG-I and MDA5 signaling by RNA viruses in innate immunity. J Virol 82:335–345

Lund JM, Alexopoulou L, Sato A, Karow M, Adams NC, Gale NW, Iwasaki A, Flavell RA (2004) Recognition of single-stranded RNA viruses by Toll-like receptor 7. Proc Natl Acad Sci USA 101:5598–5603

Lundin M, Monne M, Widell A, Von Heijne G, Persson MA (2003) Topology of the membrane-associated hepatitis C virus protein NS4B. J Virol 77:5428–5438

Meylan E, Tschopp J (2006) Toll-like receptors and RNA helicases: two parallel ways to trigger antiviral responses. Mol Cell 22:561–569

Munoz-Jordan JL, Laurent-Rolle M, Ashour J, Martinez-Sobrido L, Ashok M, Lipkin WI, Garcia-Sastre A (2005) Inhibition of alpha/beta interferon signaling by the NS4B protein of flaviviruses. J Virol 79:8004–8013

Munoz-Jordan JL, Sanchez-Burgos GG, Laurent-Rolle M, Garcia-Sastre A (2003) Inhibition of interferon signaling by dengue virus. Proc Natl Acad Sci USA 100:14333–14338

Onomoto K, Yoneyama M, Fujita T (2007) Regulation of antiviral innate immune responses by RIG-I family of RNA helicases. Curr Top Microbiol Immunol 316:193–205

Park GS, Morris KL, Hallett RG, Bloom ME, Best SM (2007) Identification of residues critical for the interferon antagonist function of Langat virus NS5 reveals a role for the RNA-dependent RNA polymerase domain. J Virol 81:6936–6946

Pichyangkul S, Endy TP, Kalayanarooj S, Nisalak A, Yongvanitchit K, Green S, Rothman AL, Ennis FA, Libraty DH (2003) A blunted blood plasmacytoid dendritic cell response to an acute systemic viral infection is associated with increased disease severity. J Immunol 171:5571–5578

Pitha PM, Kunzi MS (2007) Type I interferon: the ever unfolding story. Curr Top Microbiol Immunol 316:41–70

Qu L, McMullan LK, Rice CM (2001) Isolation and characterization of noncytopathic pestivirus mutants reveal a role for nonstructural protein NS4B in viral cytopathogenicity. J Virol 75:10651–10662

Querec T, Bennouna S, Alkan S, Laouar Y, Gorden K, Flavell R, Akira S, Ahmed R, Pulendran B (2006) Yellow fever vaccine YF-17D activates multiple dendritic cell subsets via TLR2, 7, 8 and 9 to stimulate polyvalent immunity. J Exp Med 203:413–424

Sariol CA, Munoz-Jordan JL, Abel K, Rosado LC, Pantoja P, Giavedoni L, Rodriguez IV, White LJ, Martinez M, Arana T, Kraiselburd EN (2007) Transcriptional activation of interferon-stimulated genes but not of cytokine genes after primary infection of rhesus macaques with dengue virus type 1. Clin Vaccine Immunol 14:756–766

Severa M, Fitzgerald KA (2007) TLR-mediated activation of type I IFN during antiviral immune responses: fighting the battle to win the war. Curr Top Microbiol Immunol 316:167–192

Shresta S, Sharar KL, Prigozhin DM, Beatty PR, Harris E (2006) Murine model for dengue virus-induced lethal disease with increased vascular permeability. J Virol 80:10208–10217

Shresta S, Sharar KL, Prigozhin DM, Snider HM, Beatty PR, Harris E (2005) Critical roles for both STAT1-dependent and STAT1-independent pathways in the control of primary dengue virus infection in mice. J Immunol 175:3946–3954

Takaoka A, Yanai H (2006) Interferon signaling network in innate defense. Cell Microbiol 8:907–922

Wang JP, Liu P, Latz E, Golenbock DT, Finberg RW, Libraty DH (2006) Flavivirus activation of plasmacytoid dendritic cells delineates key elements of TLR7 signaling beyond endosomal recognition. J Immunol 177:7114–7121

Warke RV, Martin KJ, Giaya K, Shaw SK, Rothman AL, Bosch I (2008) TRAIL is a novel antiviral protein against dengue virus. J Virol 82:555–564

Warke RV, Xhaja K, Martin KJ, Fournier MF, Shaw SK, Brizuela N, de Bosch N, Lapointe D, Ennis FA, Rothman AL, Bosch I (2003) Dengue virus induces novel changes in gene expression of human umbilical vein endothelial cells. J Virol 77:11822–11832

Dengue Virus Virulence and Transmission Determinants

R. Rico-Hesse

Contents

Abstract The mechanisms of dengue virus (DENV) pathogenesis are little understood because we have no models of disease; only humans develop symptoms (dengue fever, DF, or dengue hemorrhagic fever, DHF) and research has been limited to studies involving patients. DENV is very diverse: there are four antigenic groups (serotypes) and three to five genetic groups (genotypes) within each serotype. Thus, it has been difficult to evaluate the relative virulence or transmissibility of each DENV genotype; both of these factors are important determinants of epidemiology and their measurement is complex because the natural cycle of this disease involves human-mosquito-human transmission. Although epidemiological and evolutionary studies have pointed to viral factors in determining disease outcome, only recently developed models could prove the importance of specific viral genotypes in causing severe epidemics and their potential to spread to other continents. These new models involve infection of primary human cell cultures, "humanized" mice and field-collected mosquitoes; also, new mathematical

R. Rico-Hesse

Southwest Foundation for Biomedical Research, San Antonio, TX 78227, USA
e-mail: rricoh@sfbr.org

A.L. Rothman (ed.), *Dengue Virus*, Current Topics in Microbiology and Immunology 338,   45
DOI 10.1007/978-3-642-02215-9_4, © Springer-Verlag Berlin Heidelberg 2010

models can estimate the impact of viral replication, human immunity and mosquito transmission on epidemic behavior. DENV evolution does not seem to be rapid and the transmission and dispersal of stable, replication-fit genotypes has been more important in the causation of more severe epidemics. Controversy regarding viral determinants of DENV pathogenesis and epidemiology will continue until virulence and transmissibility can be measured under various conditions.

1 Introduction

Dengue virus (DENV) pathogenesis seems to be determined by numerous, interacting factors: viral virulence, host immunity and immune status, host genetics and possibly others (e.g., preexisting diseases). Because we have no models of severe dengue disease (DHF), all associations of viruses with increased pathogenesis have been indirect and painstakingly slow in being developed. Transmissibility has also been measured indirectly: the successful isolation of viruses from patients and the preponderance of one serotype over another have been documented in numerous countries but this has also introduced biases in our sampling. We do not have available a fully representative set of DENV genomes to study and understand what truly constitutes the natural range of DENV variation; many samples from mosquitoes or less-ill human infections are missing, in addition to those from countries lacking the laboratory and public health infrastructure necessary for detecting and isolating viruses. Virus-mosquito interactions also add a layer of complexity to the determination of which genotype is being transmitted and we are only beginning to measure the effects of this selection. However, many new methods have been applied to the study of DENV genetic variation, replication fitness and their effects on transmissibility and pathogenesis in humans. None of these methods are perfect and we must still regard them as surrogates for measurements of the natural viral determinants of disease. Thus, understanding this complex system will probably require multidisciplinary approaches to solving the mysteries of the interaction of the many factors that contribute to DENV epidemiology. Other, nonviral factors contributing to DENV pathogenicity and transmission are discussed in accompanying chapters.

2 Controversies Over Virulence and Evolution

The first descriptions of DENV virulence differences came from epidemiologic and entomologic studies done in the South Pacific by Rosen and Gubler, in the 1970s (Gubler et al. 1978; Rosen 1977). It was noted that some outbreaks in this region had fewer or no cases of DHF and the transmitted viruses were considered of low virulence; other outbreaks had many cases of DHF, after primary infection and these viruses were therefore more virulent. However, it took the development of RNA nucleotide sequencing techniques and the use of these sequences to generate phylogenetic trees of evolutionary relationships among viruses to discover that specific variant groups, or genotypes, were more frequently associated with dengue epidemics and severe disease (Chungue et al. 1995; Lanciotti et al. 1997; Lanciotti et al. 1994; Messer et al. 2003; Rico-Hesse 1990; Rico-Hesse et al. 1997). More recently it has been shown that some genotypes associated with DHF have been introduced and become established (endemic) in other continents, sometimes displacing the less-virulent DENV already being transmitted in those regions (causing DF only). "Virulent" genotypes have been described for serotypes 2 and 3 and it remains to be seen if further evolutionary studies will pinpoint similar groups in serotypes 1 and 4 (Rico-Hesse 2003).

In the case of DENV, there is no evidence for rapid evolution and selection as in HIV, influenza or SARS viruses; for DENV, man-made ecologic disruption or increments in the number of mosquitoes or hosts are more important than evolution towards more virulent genotypes. There has yet to be evidence for the circulation of a recombinant DENV and those recombinant genomes described to date have been the product of enzymatic amplification techniques and are thus probably lab artifacts (Aaskov et al. 2007; Holmes and Twiddy 2003; Worobey et al. 1999); no one has isolated and fully characterized a recombinant DENV that is being transmitted in nature, causing disease. Although there is evidence for recombination in other, positive-strand viruses, specific steps in DENV replication might keep this event from occurring, although ample opportunities seem to exist in multiply-infected humans and mosquitoes (Monath et al. 2005). Also, there is no evidence for the epidemic transmission of the sylvatic genotype viruses from West Africa or Malaysia. These older, seemingly less-virulent viruses are transmitted mainly by canopy-dwelling mosquitoes to monkeys and they do not cause outbreaks in the human populations inhabiting those areas. The viruses isolated from dengue patients during outbreaks belong to genotypes imported from other continents (in the case of serotype 2, a genotype originating in the Indian subcontinent was introduced to Africa) (Rico-Hesse 2003). Some researchers, with ample field experience in tropical areas, believe that these zoonotic cycles will eventually disappear, because of the constant reduction of natural forests; thus, these cycles have practically no importance as reservoirs of human dengue (Rodhain 1991).

Although we have not detected increases in replication fitness for any given genotype, some of the evolutionary events leading to more virulent strains seem to have already occurred and by the time these viruses rapidly spread from Southeast Asia to other areas of the world (1940s), they were already virulent or replicatively fit (Gubler 2002). We have yet to measure any specific genetic changes that are fixed in the viral population as virulent genotypes are successfully dispersed to other continents and we do not know whether there have been any changes imposed by selection in their new environments (e.g., for serotype 2, Southeast Asian genotype introduced into the Americas; for serotype 3, Sri Lankan genotype III introduced to the Americas). That is, these viruses had increased fitness at their origin, in that they are directly linked to the appearance of DHF and they are transmitted more efficiently by mosquitoes. However, some researchers believe differences in clinical presentation and severity of epidemics are a function of only immunologic and genetic differences between the human populations in both continents (Southeast Asia and Americas) (Halstead 2006) or that nonneutralizing antibodies formed during DENV infection play a role in gradually selecting for more pathogenic viruses in humans (Morens and Fauci 2008).

3 Controversies Over Transmission

Of special concern lately has been the effect of global warming on the incidence and spread of dengue disease. Although there are probably no effects on DENV replication in humans, if environmental temperatures rise, many investigators presume there might be an effect on virus transmission by mosquitoes. This is derived from the fact that increases in temperature (along with increases in rainfall) directly affect mosquito development (from larval to adult stages) and their populations can increase dramatically. Also, increases in average temperatures in new climes might make conditions favorable for mosquito breeding and the establishment of new populations. This could surely lead to more chances of exposure to mosquito bites for the human population and thus for infection by mosquito-borne viruses. However, the reason why many mosquito-borne diseases have yet to affect large populations in developed nations seems to be a lack of exposure to mosquito bites; air-conditioning and human behavior have been shown to reduce DENV transmission in the southern United States (Reiter et al. 2003). That is, human activities and their impact on local ecology have generally been more significant in increasing dengue prevalence; thus, dengue disease seems to be influenced more by economic than climatic factors (Gubler et al. 2001; Reiter 2001). Also, research described below has shown that increases in temperature might have more of an impact on selecting for those virulent genotypes that are already being transmitted by mosquitoes. Mosquito survival rates and the time it takes for a virus to infect and be transmitted by the mosquito seem to be more important in this context.

Another subject that has received renewed attention is the possibility of human modification or management of viral virulence by impacting transmission dynamics. This is important for the application of rapid public health measures in the event of the emergence of new strains of parasites, bacteria or viruses (Lipsitch and Moxon 1997). We presume we can influence virulence by the application of changes in human habits or by the use of control factors such as vaccines – and there are numerous precedents of how public health measures have changed the population dynamics of microorganisms (e.g., there are now more cases of vaccine-induced polio or yellow fever in some countries). Efforts to understand the relationship between parasite adaptation to hosts, virulence and transmission have developed into a small industry in evolutionary biology (Bull and Dykhuizen 2003). Although most discussion is still theoretical, the relationships between virulence and transmission have been weighed, presuming that there is an evolutionary trade-off for optimizing either factor within an organism (Ebert and Bull 2003). That is, to increase its chances of transmission to another host, an organism will limit its replication or virulence in so far as to not kill its host. For example, highly virulent, zoonotic viruses such as Ebola and rabies are less transmissible than measles or common cold viruses, which rarely kill their human host. However, this is overly simplistic when it comes to the evolution of viruses that have multiple strains (multiple infections, with some cross-protection), where virulence involves immunopathology, or where there is another host or an amplifying vector involved in transmission (Day et al. 2007). Thus, the main goal for disease control is to understand how we can reduce virulence in a virus population without making its level of transmission higher. However, this seems a daunting task if we include the effects of evolution by individual versus group selection, bottlenecks and changes in fitness trade-offs. The concern here is whether we will shift DENV evolution and population dynamics by applying incomplete control strategies (e.g., nonsterilizing vaccination or vector transformation).

4 Methodologic Advances and Findings

The controversies mentioned above point to factors we should consider or understand in the control of dengue disease: evolutionary studies of DENV have helped us concentrate on detection (diagnosis and sampling), analysis (genetic and phenotype) and control (transmission dynamics in host and mosquitoes) of those genotypes already shown to be the culprits (Rico-Hesse 2007). The methods described below could help us reach these goals.

4.1 Rapid Sequence Analysis

The determination of viral RNA sequences from different areas of the genome has now become routine and numerous laboratories around the world have this capability; this has added exponentially to the number of DENV samples available for comparison in GenBank. The comparison of these nucleotides and their encoded amino acids can be done with sophisticated computer algorithms that can tell us much about the rates and sites of mutation or evolution in the viral genome. These data can then be matched with patient viral loads, diagnoses, outbreak characteristics and transmission distribution, to look for specific associations. The RT-PCR technique has allowed for the enzymatic amplification of these sequences from very small amounts of viral RNA from almost any type of tissue but many researchers have now avoided virus isolation and characterization, thus introducing mistakes in some of the banked information (e.g., from amplification, sequencing, or cloning artifacts) and without information on viability or antigenicity. Therefore, it is also important to have access to classic virology techniques, especially if one is to derive information about virus phenotype.

The comparisons of full genome sequences of many DENV have helped pinpoint differences that could be involved in virulence: the comparison of viruses from two different genotypes associated with DHF or DF only (Southeast Asia and Americas, respectively) showed that there were consistent differences in the 5′-untranslated region (UTR), one envelope protein site (aa390) and the 3′-UTR of the DENV serotype 2 genome (Leitmeyer et al. 1999). Comparisons within the Southeast Asian genotype did not identify any specific nucleotides associated with producing DHF (Mangada and Igarashi 1998; Pandey and Igarashi 2000), so we assume all viruses of this genotype have the potential to produce severe disease. This has also been the case with other DENV, where recent studies have suggested that differences in 5′- and 3′-UTR in the genome can alter levels of replication (Miagostovich et al. 2006; Sirigulpanit et al. 2007; Tajima et al. 2007), which can be extrapolated to viral load in blood or disease presentation (Wang et al. 2006). Another chapter in this volume describes how these influences may occur.

4.2   Infection of Primary Human Cells

The first targets of DENV replication, after mosquito bite, were postulated to be monocytes or macrophages and numerous studies focused on these cell types. However, more recent studies, using newer technologies for cell identification (mainly flow cytometry) have shown that DENV infects human monocytes poorly compared to dendritic cells, including Langerhans cells and monocyte-derived dendritic cells (Marovich et al. 2001; Wu et al. 2000). Primary dendritic cell cultures can be derived from human peripheral blood donations to banks and this is the usual source for studies to compare the replication of low-passage DENV from patients. Because these samples are obtained anonymously, we must be careful to obtain them from blood banks in areas where there is no DENV transmission (i.e., no antigenic priming of cells) and we cannot determine the human genetic background that might lead to differences in virus replication. However, studies reported in 2003 and 2005 (Cologna et al. 2005; Cologna and Rico-Hesse 2003) were able to show consistent differences in replication of DENV of two genotypes of serotype 2, demonstrating that there is an ex vivo correlation to the virulent phenotype derived from evolutionary studies. Although there seem to be innate, probably genetic differences in the yields of virus produced by cells from individual donors, this variation could be accounted for statistically and the corre- lations with virulence of patient-derived viruses were established (and note that these differences occur in the absence of antibodies). These primary cell cultures were also used to test recombinant viruses, to determine the influence of specific genome regions on virus replication and yields from human cell targets. These studies confirmed that the exchange of three genomic regions (5′ and 3′ UTRs, and E390) could reduce the levels of replication and virus yields of a Southeast Asian virus to those of wild-type, less virulent viruses of the American genotype (Cologna and Rico-Hesse 2003). Other uses for cultured primary human cells include the identification of specific cells that are producing more virus (tropism) and whether virus replication is even required for pathologic effects.

4.3   Infection of Field-Collected Mosquitoes

Another step in determining if a virulent genotype had an increased transmission fitness phenotype was to study differences in replication and dissemination (i.e., the possibility of transmitting virus by bite) in the natural mosquito vector, *Aedes aegypti*. Laboratory-reared colonies of mosquitoes (e.g., Rexville or Rockefeller strains) seem to have lost their selectivity for infection and it is recommended that mosquitoes used in these experiments be from the F4 generation or lower (F0 = field-collected eggs). The virulent, Southeast Asian strains of serotype 2 were shown to infect a larger proportion of mosquitoes than the less virulent, American genotype strains, after feeding mosquitoes on blood containing the same titer of virus (Armstrong and Rico-Hesse 2001); also, a greater proportion of mosquitoes develop disseminated infections with the virulent genotype (Armstrong and Rico-Hesse 2003). If mosquitoes were fed both genotypes simultaneously, they were much more likely (sevenfold) to develop an infection with the virulent strains (Cologna et al. 2005). When the dynamics of virus replication and dissemination were compared for both genotypes, the virulent strains had reached the salivary glands up to 7 days earlier than the less virulent viruses (Anderson and Rico-Hesse 2006). This means that virulent strains may replicate and be transmitted much sooner to human hosts, outcompeting the less virulent viruses and causing many more cases of disease, thus ecologically displacing those that cause less severe dengue. This efficiency of transmission by the vector could explain how certain genotypes have displaced others, shifting the evolution of dengue disease towards more virulence (i.e., more DHF).

4.4   Mouse Models of Disease

During this decade, major advances have been made in the development of new mouse breeds and their transplantation with human stem cells (from umbilical cord blood cells) that may effectively mimic the human immune system or show human signs of disease upon infection. The combination of studies in these mice with those in mice that are defective in interferon production or receptors have led to insights into the mechanisms of dengue pathogenesis (Bente et al. 2005; Kuruvilla et al. 2007; Kyle et al. 2007; Shresta et al. 2006). Thus far, none of these models develop DHF and the production of DENV-specific antibodies has been low or undetectable. However, others have shown that mice engrafted with human hematopoietic cells can be effectively used to study pathogenesis of viruses for which no other models exist (Melkus et al. 2006; Watanabe et al. 2007). It is anticipated that after adaptation of this system to DENV infection by multiple strains, with the acquisition of DENV-specific, functional B and T cells, that the signs of DHF might appear in these "humanized" mice. This would finally allow for the measurement of the many effects of immunopathogenesis, including the protective cross-immunity created by serial infection and the evaluation of many basic questions, such as dose-dependence of infection, the relevance of mosquito factors to infection (e.g., salivary gland proteins) and the role of other cells as primary targets of infection (e.g., endothelial cells). Most importantly, this model could allow for the immediate testing of antivirals and vaccine candidates, where effective systems for testing products before human use have been lacking.

4.5 *Mathematical Models of Transmission*

Another promising new field has been the development of mathematical models of DENV transmission, including "evolutionary epidemiology" and virulence management. These models are being used to estimate the effect of changing host immunity or mosquito transmission on the amount of virus circulating and the risks to a hypothetical human population and epidemic "topology." These analyzes have suggested that DENV cross-serotype immunity and mosquito demographics, rather than immune enhancement, are the most important determinants in the dynamics of specific serotype cycles or genotype replacement during epidemics and that the application of incomplete control strategies might actually increase the incidence of severe disease (Adams et al. 2006; Cummings et al. 2005; Nagao and Koelle 2008; Wearing and Rohani 2006). Although these models are complex and still require many basic measurements for their refinement, some of the details are being added as they become available from laboratory or ecological studies (e.g., quantification of cross-protection by various DENV strains and many different measurements of mosquito transmission dynamics, including vector genetics and their effect on competence and capacity, etc.). Other applications of computer modeling have involved measuring the importance of host genetics over parasite contributions to virulence. For virology, the importance of host genetics in disease pathogenesis has been discussed for many years but recent technologies have allowed researchers to weigh the influence of virus virulence over human host genetics, in the case of pandemic influenza A (Gottfredsson et al. 2008; Pitzer et al. 2007). However, these approaches are extremely controversial at this point, as other investigators have reached opposing conclusions when using similar methods and results (Albright et al. 2008).

5 Perspective

Only recently have public health officials in developed countries become concerned about the increased transmission and geographic spread of DENV. In many cases this is due to the increase in cases imported by tourists from less-developed tropical regions. For the United States, there has been a marked increase in imported cases and autochthonous transmission in Texas and Hawaii during this decade (Morens and Fauci 2008) and for the first time we have been able to document the introduction of a virulent DENV genotype into this country (CDC 2007; Rico-Hesse 2007). This has added a sense of urgency to research using some of the models described here and a hope for additional support for studies of this long-neglected tropical disease.

Acknowledgments Financial support to R.R.-H. was provided by the NIH (AI50123) and the Robert J. and Helen C. Kleberg Foundation.

References

Aaskov J, Buzacott K, Field E, Lowry K, Berlioz-Arthaud A, Holmes EC (2007) Multiple recombinant dengue type 1 viruses in an isolate from a dengue patient. J Gen Virol 88:3334–3340

Adams B, Holmes EC, Zhang C, Mammen MP Jr, Nimmannitya S, Kalayanarooj S, Boots M (2006) Cross-protective immunity can account for the alternating epidemic pattern of dengue virus serotypes circulating in Bangkok. Proc Natl Acad Sci USA 103:14234–14239

Albright FS, Orlando P, Pavia AT, Jackson GG, Cannon Albright LA (2008) Evidence for a heritable predisposition to death due to influenza. J Infect Dis 197:18–24

Anderson JR, Rico-Hesse R (2006) *Aedes aegypti* vectorial capacity is determined by the infecting genotype of dengue virus. Am J Trop Med Hyg 75:886–892

Armstrong PM, Rico-Hesse R (2001) Differential susceptibility of *Aedes aegypti* to infection by the American and Southeast Asian genotypes of dengue type 2 virus. Vector Borne Zoonotic Dis 1:159–168

Armstrong PM, Rico-Hesse R (2003) Efficiency of dengue serotype 2 virus strains to infect and disseminate in *Aedes aegypti*. Am J Trop Med Hyg 68:539–544

Bente DA, Melkus MW, Garcia JV, Rico-Hesse R (2005) Dengue fever in humanized NOD/SCID mice. J Virol 79:13797–13799

Bull J, Dykhuizen D (2003) Virus evolution: epidemics-in-waiting. Nature 426:609–610

CDC (2007) Dengue hemorrhagic fever–U.S.-Mexico border, 2005. MMWR 56:785–789

Chungue E, Cassar O, Drouet MT, Guzman MG, Laille M, Rosen L, Deubel V (1995) Molecular epidemiology of dengue-1 and dengue-4 viruses. J Gen Virol 76:1877–1884

Cologna R, Armstrong PM, Rico-Hesse R (2005) Selection for virulent dengue viruses occurs in humans and mosquitoes. J Virol 79:853–859

Cologna R, Rico-Hesse R (2003) American genotype structures decrease dengue virus output from human monocytes and dendritic cells. J Virol 77:3929–3938

Cummings DA, Schwartz IB, Billings L, Shaw LB, Burke DS (2005) Dynamic effects of antibody-dependent enhancement on the fitness of viruses. Proc Natl Acad Sci USA 102:15259–15264

Day T, Graham AL, Read AF (2007) Evolution of parasite virulence when host responses cause disease. Proc Biol Sci 274:2685–2692

Ebert D, Bull JJ (2003) Challenging the trade-off model for the evolution of virulence: is virulence management feasible? Trends Microbiol 11:15–20

Gottfredsson M, Halldorsson BV, Jonsson S, Kristjansson M, Kristjansson K, Kristinsson KG, Love A, Blondal T, Viboud C, Thorvaldsson S et al (2008) Lessons from the past: Familial aggregation analysis of fatal pandemic influenza (Spanish flu) in Iceland in 1918. Proc Natl Acad Sci USA 105:1303–1308

Gubler DJ (2002) Epidemic dengue/dengue hemorrhagic fever as a public health, social and economic problem in the 21st century. Trends Microbiol 10:100–103

Gubler DJ, Reed D, Rosen L, Hitchcock JR Jr (1978) Epidemiologic, clinical and virologic observations on dengue in the Kingdom of Tonga. Am J Trop Med Hyg 27:581–589

Gubler DJ, Reiter P, Ebi KL, Yap W, Nasci R, Patz JA (2001) Climate variability and change in the United States: potential impacts on vector- and rodent-borne diseases. Environ Health Perspect 109:223–233

Halstead SB (2006) Dengue in the Americas and Southeast Asia: do they differ? Rev Panam Salud Publica 20:407–415

Holmes EC, Twiddy SS (2003) The origin, emergence and evolutionary genetics of dengue virus. Infect Genet Evol 3:19–28

Kuruvilla JG, Troyer RM, Devi S, Akkina R (2007) Dengue virus infection and immune response in humanized RAG2(−/−) gamma(c)(−/−) (RAG-hu) mice. Virology 369:143–152

Kyle JL, Beatty PR, Harris E (2007) Dengue virus infects macrophages and dendritic cells in a mouse model of infection. J Inf Dis 195:1808–1817

Lanciotti RS, Gubler DJ, Trent DW (1997) Molecular evolution and phylogeny of dengue-4 viruses. J Gen Virol 78:2279–2284

Lanciotti RS, Lewis JG, Gubler DJ, Trent DW (1994) Molecular evolution and epidemiology of dengue-3 viruses. J Gen Virol 75:65–75

Leitmeyer KC, Vaughn DW, Watts DM, Salas R, Villalobos I, De C, Ramos C, Rico-Hesse R (1999) Dengue virus structural differences that correlate with pathogenesis. J Virol 73:4738–4747

Lipsitch M, Moxon ER (1997) Virulence and transmissibility of pathogens: what is the relationship? Trends Microbiol 5:31–37

Mangada MN, Igarashi A (1998) Molecular and in vitro analysis of eight dengue type 2 viruses isolated from patients exhibiting different disease severities. Virology 244:458–466

Marovich M, Grouard-Vogel G, Louder M, Eller M, Sun W, Wu SJ, Putvatana R, Murphy G, Tassaneetrithep B, Burgess T et al (2001) Human dendritic cells as targets of dengue virus infection. J Investig Dermatol Symp Proc 6:219–224

Melkus MW, Estes JD, Padgett-Thomas A, Gatlin J, Denton PW, Othieno FA, Wege AK, Haase AT, Garcia JV (2006) Humanized mice mount specific adaptive and innate immune responses to EBV and TSST-1. Nature Med 12:1316–1322

Messer WB, Gubler DJ, Harris E, Sivananthan K, de Silva AM (2003) Emergence and global spread of a dengue serotype 3, subtype III virus. Emerg Infect Dis 9:800–809

Miagostovich MP, dos Santos FB, Fumian TM, Guimaraes FR, da Costa EV, Tavares FN, Coelho JO, Nogueira RM (2006) Complete genetic characterization of a Brazilian dengue virus type 3 strain isolated from a fatal outcome. Mem Inst Oswaldo Cruz 101:307–313

Monath TP, Kanesa-Thasan N, Guirakhoo F, Pugachev K, Almond J, Lang J, Quentin-Millet MJ, Barrett AD, Brinton MA, Cetron MS et al (2005) Recombination and flavivirus vaccines: a commentary. Vaccine 23:2956–2958

Morens DM, Fauci AS (2008) Dengue and hemorrhagic fever: a potential threat to public health in the United States. JAMA 299:214–216

Nagao Y, Koelle K (2008) Decreases in dengue transmission may act to increase the incidence of dengue hemorrhagic fever. Proc Natl Acad Sci USA 105:2238–2243

Pandey BD, Igarashi A (2000) Severity-related molecular differences among nineteen strains of dengue type 2 viruses. Microbiol Immunol 44:179–188

Pitzer VE, Olsen SJ, Bergstrom CT, Dowell SF, Lipsitch M (2007) Little evidence for genetic susceptibility to influenza A (H5N1) from family clustering data. Emerg Infect Dis 13:1074–1076

Reiter P (2001) Climate change and mosquito-borne disease. Environ Health Perspect 109 (Suppl 1):141–161

Reiter P, Lathrop S, Bunning M, Biggerstaff B, Singer D, Tiwari T, Baber L, Amador M, Thirion J, Hayes J et al (2003) Texas lifestyle limits transmission of dengue virus. Emerg Infect Dis 9:86–89

Rico-Hesse R (1990) Molecular evolution and distribution of dengue viruses type 1 and 2 in nature. Virology 174:479–493

Rico-Hesse R (2003) Microevolution and virulence of dengue viruses. Adv Virus Res 59:315–341

Rico-Hesse R (2007) Dengue virus evolution and virulence models. Clin Infect Dis 44:1462–1466

Rico-Hesse R, Harrison LM, Salas RA, Tovar D, Nisalak A, Ramos C, Boshell J, de Mesa MT, Nogueira RM, da Rosa AT (1997) Origins of dengue type 2 viruses associated with increased pathogenicity in the Americas. Virology 230:244–251

Rodhain F (1991) The role of monkeys in the biology of dengue and yellow fever. Comp Immunol Microbiol Infect Dis 14:9–19

Rosen L (1977) The Emperor's New Clothes revisited, or reflections on the pathogenesis of dengue hemorrhagic fever. Am J Trop Med Hyg 26:337–343

Shresta S, Sharar KL, Prigozhin DM, Beatty PR, Harris E (2006) Murine model for dengue virus-induced lethal disease with increased vascular permeability. J Virol 80:10208–10217

Sirigulpanit W, Kinney RM, Leardkamolkarn V (2007) Substitution or deletion mutations between nt 54 and 70 in the 5' non-coding region of dengue type 2 virus produce variable effects on virus viability. J Gen Virol 88:1748–1752

Tajima S, Nukui Y, Takasaki T, Kurane I (2007) Characterization of the variable region in the 3' non-translated region of dengue type 1 virus. J Gen Virol 88:2214–2222

Wang WK, Chen HL, Yang CF, Hsieh SC, Juan CC, Chang SM, Yu CC, Lin LH, Huang JH, King CC (2006) Slower rates of clearance of viral load and virus-containing immune complexes in patients with dengue hemorrhagic fever. Clin Infect Dis 43:1023–1030

Watanabe S, Ohta S, Yajima M, Terashima K, Ito M, Mugishima H, Fujiwara S, Shimizu K, Honda M, Shimizu N et al (2007) Humanized NOD/SCID/IL2Rgamma(null) mice transplanted with hematopoietic stem cells under nonmyeloablative conditions show prolonged life spans and allow detailed analysis of human immunodeficiency virus type 1 pathogenesis. J Virol 81:13259–13264

Wearing HJ, Rohani P (2006) Ecological and immunological determinants of dengue epidemics. Proc Natl Acad Sci USA 103:11802–11807

Worobey M, Rambaut A, Holmes EC (1999) Widespread intra-serotype recombination in natural populations of dengue virus. Proc Natl Acad Sci USA 96:7352–7357

Wu SJ, Grouard-Vogel G, Sun W, Mascola JR, Brachtel E, Putvatana R, Louder MK, Filgueira L, Marovich MA, Wong HK et al (2000) Human skin Langerhans cells are targets of dengue virus infection. Nature Med 6:816–820

Systemic Vascular Leakage Associated with Dengue Infections – The Clinical Perspective

Dinh T. Trung and Bridget Wills

Contents

Abstract Vascular leakage is the most serious complication of dengue infection. However, despite considerable progress in understanding the immunological derangements associated with dengue, the pathogenic mechanisms underlying the change in vascular permeability remain unclear. Lack of suitable model systems that manifest permeability characteristics similar to human vascular endothelium has seriously impeded research in this area. Similarly, limited knowledge of the factors regulating intrinsic microvascular permeability in health, together with limited understanding of the alterations seen in disease states in general, has also hampered progress. Fortunately considerable advances have been made in the field of endothelial biology in recent years, especially following appreciation of the crucial role played by the endothelial surface glycocalyx, acting in concert with underlying cellular structures, in regulating fluid flow across the microvasculature. We review what is known about vascular leakage during dengue infections, particularly in relation to current knowledge of vascular physiology, and discuss potential areas of research that may help to elucidate the complex nature of this singular phenomenon in the future.

D.T. Trung and B. Wills (✉)

Oxford University Clinical Research Unit, Hospital for Tropical Diseases, 190 Ben Ham Tu, Quan 5, Ho Chi Minh City, Vietnam
e-mail: bwills@oucru.org

A.L. Rothman (ed.), *Dengue Virus*, Current Topics in Microbiology and Immunology 338,   57
DOI 10.1007/978-3-642-02215-9_5, © Springer-Verlag Berlin Heidelberg 2010

1   Introduction

In the late 1950s astute clinicians grappling with a new and serious clinical syndrome associated with dengue infection identified plasma leakage to be one of the major features of the syndrome. (Cohen and Halstead 1966; Nimmannitya et al. 1969) Although other manifestations of severe dengue infection are recognised (profound thrombocytopenia, coagulopathy, hepatic failure etc), it is clear that the critical determinant of overall disease severity in the majority of cases is hypovolaemia, secondary to increased systemic vascular permeability and plasma leakage. Fifty years have passed since the importance of this phenomenon was first recognised and yet the underlying molecular mechanisms continue to elude us, despite considerable advances in our understanding of dengue pathogenesis in general. Prevention of dengue infections is clearly the ultimate goal but, until such time as widespread deployment of safe and effective vaccines throughout endemic areas becomes a realistic possibility, health care professionals must continue to focus on providing the best available care for symptomatic patients. It is to be hoped that improved understanding of the determinants and specific characteristics of dengue associated vascular leakage will allow informed decisions to be made regarding clinical management of hypovolaemia and shock. In addition, knowledge of the underlying mechanisms controlling leakage may allow for the design of precise pharmacological interventions to counteract or prevent it in high risk groups.

As detailed elsewhere in this book, many factors are now known to influence the severity of vascular leakage. It is clear that infection with one dengue serotype elicits immunity to that serotype but does not provide long-term cross-protective immunity to the remaining serotypes and that severe disease occurs predominantly in patients experiencing a second or subsequent infection. The generally accepted "antibody-dependent enhancement" hypothesis suggests that residual heterotypic non-neutralising antibodies from the earlier infection bind to the new virus and amplify viral replication, and the resulting increase in viral load then drives an immunopathogenic cascade that alters microvascular structure or function in some way, thereby resulting in a transient increase in permeability. Rapid mobilisation of serotype cross-reactive memory T cells has been proposed as an alternative mechanism to trigger the inflammatory cascade and many other factors such as differences in viral virulence, molecular mimicry, immune complex and/or complement mediated dysregulation, genetic predisposition etc. have been shown to correlate with disease severity. However, as yet no mechanism has been identified that links any of these established immunological derangements with a definitive effect on microvascular structure or function consistent with the observed alteration in permeability. In addition, most of the immunological abnormalities so far identified do not differ substantially from those seen in other infections without an apparent effect on permeability.

One reason for the lack of progress has been the inevitable focus of much dengue research on immunological aspects of disease or on host/pathogen interactions, reflecting the critical importance of understanding these pathways for successful vaccine design. Research on microvascular dysfunction per se has been limited, at least in part due to the rather crude understanding of the mechanisms controlling microvascular permeability that has prevailed until recently and also to the limited range of tools available to interrogate this phenomenon in vivo. Undoubtedly progress has also been hampered by the lack of a robust animal model demonstrating susceptibility to vascular leakage and reflecting the human disease phenotype; recently a murine model exhibiting altered permeability has been described and may help to provide mechanistic insights into in vivo disease pathogenesis in the future. (Shresta et al. 2006) Various groups have developed in vitro model systems to look at endothelial cell function in response to dengue infection and/or to examine the effects of different immune mediators. (Bunyaratvej et al. 1997; Avirutnan et al. 1998; Bonner and O'Sullivan 1998; Lin et al. 2003) However, even the most sophisticated model systems still approximate poorly to the permeability characteristics of vascular endothelium in vivo and the true relevance of such systems to the human disease process is doubtful. (Potter and Damiano 2008) Fortunately, in recent years considerable advances have been made in elucidating the normal physiology of the microvasculature, (Michel and Curry 1999; Weinbaum et al. 2007) together with some progress in understanding the derangements that are seen in systemic disorders with prominent microvascular complications such as diabetes and hypertension. (van den Berg et al. 2003; Perrin et al. 2007)

In this chapter we will review what is known from a clinical perspective about vascular leakage during dengue infections, particularly in relation to current knowledge of vascular physiology and we will discuss potential areas of research that might help to elucidate the complex nature of this singular phenomenon in the future.

2 The Timing and Evolution of Dengue Associated Vascular Leakage

Information on the onset, evolution and duration of the vascular leak process is limited, largely due to technical difficulties in measuring capillary ultrafiltration and identifying increased permeability. It is frequently stated that a sudden onset of severe leakage occurs at or shortly after defervescence, during the critical phase of the illness. Whilst it is true that this is often the time at which leakage becomes clinically apparent by virtue of its effects on intravascular/interstitial fluid balance, there are strong indications that the alteration in systemic permeability begins earlier in the evolution of the disease.

Normally, plasma volume is regulated within tightly circumscribed limits by complex homeostatic mechanisms. (Guyton and Hall 2000) Although catastrophic leak occurring over a short period of time – as seen in meningococcal septicaemia – may overwhelm these mechanisms, such severe leak is invariably accompanied by cardiovascular collapse within hours. Less dramatic increases in ultrafiltration result in substantial increases in lymphatic flow (up to ten-fold) which, together with renal and adrenal compensatory mechanisms, serve to maintain plasma volume close to normal. Pleural effusions, ascites and other overt signs of leakage do not become apparent until these mechanisms are significantly compromised. Unfortunately, lymphatic flow is difficult to measure clinically – direct quantification of changes in microvascular fluid flow is thus rarely possible and we are left with observing the secondary effects on plasma volume and body fluid distribution.

Shock is actually relatively infrequent in patients with dengue, with many patients manifesting clear evidence of leakage without cardiovascular compromise. This clinical pattern supports the idea of a relatively slow sustained process that allows time for activation of the homeostatic regulatory mechanisms, rather than a sudden massive leak; only in patients in whom the leak exceeds the regulatory capacity is cardiovascular compromise likely to become apparent. One early study performed in the 1960s investigating transvascular escape of labelled human albumin demonstrated that leakage was indeed present during the febrile phase of the illness, becoming more severe later when hypovolaemic shock developed. (Suwanik et al. 1967) Currently, the most common method of monitoring leakage in dengue patients relies on identification of relative haemoconcentration, determined by tracking increases in serial haematocrit measurements. However, the method is rather insensitive and suffers from the serious limitation that an individual's baseline value is rarely known. In addition most published series focus primarily on increases during the critical period, when hypovolaemic shock is most likely to be seen, and there is little information on early haematocrit changes. Recently a body of work has emerged utilising serial ultrasound examinations to diagnose dengue associated plasma leakage. (Setiawan et al. 1995; Thulkar et al. 2000; Wu et al. 2004; Venkata Sai et al. 2005; Colbert et al. 2007; Srikiatkhachorn et al. 2007) As expected, effusions, ascites and gallbladder wall oedema (a nonspecific marker of leakage) are commonly found around the time of defervescence and correlate with disease severity. However evidence of minor leakage has also been demonstrated as early as day 2–3 of fever in patients with relatively mild disease. In addition, in a recent study of artificially induced dengue infections in healthy volunteers with no previous flavivirus exposure, more than half of the infected subjects showed ultrasound evidence of sub-clinical fluid accumulation. (Statler et al. 2008)

Whilst not measuring leakage directly, strain gauge plethysmography is a noninvasive technique that can be employed to measure the potential for leakage in human subjects. (Gamble et al. 1993) In essence the technique measures the filtration capacity of the microvasculature, giving a numerical estimate of the capillary filtration coefficient (K_f), a theoretical entity dependent on the physical properties and surface area of the capillary bed. Starling originally hypothesised in 1896 that opposing hydrostatic and oncotic pressures in the capillaries of the microcirculation governed fluid flow (J_v) between the intravascular and interstitial body fluid compartments. (Starling 1896) Although he never expressed his ideas in mathematical terms, his basic hypothesis, refined and developed by others in the twentieth century, has become accepted as the cornerstone of modern thinking about fluid exchange in the microcirculation and is often expressed in the form of the Starling equation, $J_v - K_f [(P_c - P_i) - \sigma (\pi_c - \pi_i)]$. (Michel 1997) It is known that K_f is age dependent, with higher values demonstrated in children than adults, (Gamble et al. 2000) and this may be one reason why children with dengue develop hypovolaemic shock more readily than adults. Young mammals have a larger microvascular surface area per unit volume of skeletal muscle than adults and, in addition, developing micro-vessels are known to be more permeable to water and plasma proteins than when mature. A study in Vietnam indicated that in children with dengue and vascular leakage, the capillary filtration coefficient was approximately 50% higher than that for control subjects of a similar age but there was no significant difference between patients with and without shock. (Bethell et al. 2001) It was also apparent that minor abnormalities persisted after resolution of clinical symptoms. Further research in this area has been hampered by the delicate nature of the strain gauge apparatus, which requires considerable skill and patience to give reproducible results; however, with the modern instruments now available there is potential for new studies, for example examining the timing of onset and duration of vascular leak in relation to clinical disease manifestations, investigating whether patients with mild disease exhibit real increases in ultrafiltration and exploring possible relationships with viral and/ or immunological parameters.

3   Microvascular Permeability in Health

Although the clinical features of severe dengue indicate that vascular leakage is a prominent feature of the disease and the dengue virus is known to infect cells of endothelial derivation in vitro, there is, as yet, no convincing evidence for endothelial cell infection in vivo. A small number of histopathological studies have been performed on human tissue but only minor non-specific changes have been demonstrated in the microvasculature. (Sahaphong et al. 1980; Jessie et al. 2004; Balsitis et al. 2009; de Araujo et al. 2009) How can this apparent contradiction be explained?

To try to understand the phenomenon we must examine the mechanisms operating at the endothelial barrier in the resting state. Within the extracellular fluid compartment, plasma and interstitial fluid exist in dynamic equilibrium separated by the semi-permeable capillary wall consisting of endothelial cells and their associated surface glycocalyx layer. (Michel and Curry 1999) This highly anionic fibre matrix of proteoglycans, glycosaminoglycans (GAGs) and adherent plasma proteins is located on the luminal surface of the vascular endothelium, anchored in the plasma membrane of the endothelial cells. The glycocalyx is now considered to be the primary barrier to the movement of water and molecules, effectively functioning as a molecular sieve to selectively restrict molecules within the circulating plasma and thus limit access to underlying cellular transport mechanisms. The endothelial cells form a secondary barrier, with fluid flow thought to occur principally through the inter-endothelial gap junctions via openings that are too large to account independently for the known molecular sieving properties of the vascular wall. (Adamson and Michel 1993) Molecular size, configuration and charge all play a role in determining the disposition and movement of macromolecules within the system; thus small solutes are freely filtered, the clearance of larger molecules decreases with increasing size and those with an effective molecular radius of more than approximately 4.2 nm are almost entirely restricted within plasma. Albumin, the major protein responsible for the colloidal properties of plasma, has a molecular radius of 3.6 nm but carries a strong negative charge and is filtered less readily than neutral proteins of similar size.

Most studies attempting to look at systemic vascular leak mechanisms have focused on changes at the endothelial cell level (i.e., the secondary barrier) in response to vasoactive mediators, rather than at changes in intrinsic permeability involving the primary glycocalyx barrier. Using isolated perfused single-vessel preparations in small mammals, many cytokines and other chemical mediators have been shown to induce gap formation in vascular endothelium, usually in post-capillary venules rather than true capillaries. (Majno and Palade 1961) Evidence from electron microscopy (EM) studies suggests that these gaps are mainly intercellular, probably caused by shrinkage of adjacent cells away from one another, although some transcellular openings have also been documented. (Michel and Neal 1999) Different mediators appear to open different pathways, (Schaeffer et al. 1993; Michel and Kendall 1997) but in general the responses are rapid, transient and associated with overall loss of selectivity to macromolecules. In addition, cellular components of blood may pass through intercellular pathways together with plasma constituents.

Most of the in vivo information pertaining to vascular leak syndromes in humans relates to glomerular capillary leak, since disorders such as nephrotic syndrome are relatively common and urine is an easily accessible fluid to study. Knowledge of the sieving properties of the glomerular capillaries can be obtained by measuring the fractional clearance of test macromolecules. (Myers and Guasch 1993; Blouch et al. 1997) The clearance of a specific molecule is equal to the excretion in the urine per unit time divided by the concentration in plasma, provided there is no tubular modification of the solute in question. Relating this clearance to that of a freely filtered reference marker that is neither secreted nor absorbed adjusts for the glomerular filtration rate. Since many endogenous proteins are either secreted or absorbed in the tubules, fractional clearance measurements of non-reabsorbable polymers with variable size and charge characteristics, usually dextrans or Ficoll, are preferred for research. Direct comparison of plasma and interstitial fluid protein concentrations is not generally feasible in human studies but, given the highly specialised renal adaptations to prevent protein loss from the very extensive renal vascular bed, any increase in glomerular protein permeability demonstrated in patients without renal disease is likely to indicate a substantial increase in systemic vascular permeability.

Based on this premise, measurement of fractional urinary protein clearances has been used to provide valuable insights into the pathophysiology of the systemic vascular leak associated with meningococcal septicaemia. (Oragui et al. 2000)

Increased urinary clearances of albumin and IgG were demonstrated in children with meningococcal sepsis relative to controls, with proportionately greater losses of albumin (MW 69 kD) than IgG (MW 150 kD) and a clear relationship to disease severity. The leak was associated with increased plasma and urine concentrations of GAGs. The authors postulate that the correlation between the severity of protein leakage and the urinary excretion of GAGs indicates that the loss of GAGs may be causally related to the increase in permeability to proteins – i.e., that cleavage of GAGs from the luminal surface of the vascular endothelium may result in loss of endothelial negative charge and consequent leakage of proteins. This is supported by experimental data showing that endothelial GAGs can be degraded and surface charge and GAG activity altered, by inflammatory mediators and neutrophils.

4   The Characteristics of Dengue Associated Vascular Leakage

Assessment of the characteristics of the protein leak seen in patients with dengue should provide useful insights into the underlying pathophysiology. Hypoalbuminaemia is well documented and correlates with severity. Proteinuria is rarely assessed but has been reported in a few case series. In one study hypoalbuminaemia and proteinuria were noted in 28% and 22% respectively of dengue-infected patients, in many within the first few days from the onset of fever. (Garcia et al. 1995) Fractional clearance estimations for endogenous proteins have demonstrated that smaller proteins such as albumin and transferrin are relatively more affected than IgG – indicating that size selectivity is at least partially retained during acute infections – and support the view that charge selectivity may be impaired. (Wills et al. 2004) In the same study urinary GAG excretion was significantly increased in children with DSS, suggesting a possible role for disruption of the surface glycocalyx in the pathogenesis of the vascular leak. Studies using dextran clearance methodology to examine size and charge selectivity in more detail should prove instructive.

   Novel techniques that have been recently developed to explore systemic microvascular dysfunction in other disease states should also be considered. There is increasing interest in the role of damage to the endothelial cell/glycocalyx complex in the pathogenesis of various chronic progressive disorders such as arteriosclerosis, hypertension and diabetes, all of which are thought to be associated with a slight increase in the transcapillary filtration rate although the mechanisms remain unknown. There is good evidence to support the presence of an increase in vascular permeability in diabetics prior to the development of overt microangiopathy, thought to be due to an as yet unexplained effect of hyperglycemia on the surface glycocalyx layer. (Perrin et al. 2007) In attempting to understand this phenomenon a group in Amsterdam has developed a novel method to estimate the total volume of the glycocalyx in the body; the method is based on the premise that there are two distribution compartments within the intravascular space, the circulating plasma and a non-circulating component located within the endothelial surface layer. (Vink and Duling 2000; Michel and Curry 2009) Different tracer molecules permeate this layer at different rates. The group has demonstrated a significant reduction in glycocalyx volume in diabetics compared to controls, consistent with generalised thinning of the layer. A modification of the Dutch technique is being used by our group in Vietnam to estimate glycocalyx volume during acute dengue infections compared to similar measurements made after recovery. Depending on the findings, this may also prove to be a useful technique to look in more detail at the natural history of the altered permeability since repeated assessments can be made over time.

While this approach may yield interesting information regarding possible structural changes in the surface glycocalyx, what is really needed is a more direct approach. Unfortunately, since the matrix is highly hydrated in vivo, it dehydrates rapidly in tissue specimens and undergoes shrinkage unless carefully preserved. Perfusion with supravital cationic dyes prior to biopsy allows excellent visualisation on EM (van den Berg et al. 2003) but since such dyes are highly toxic use in human studies is contra-indicated. However, single microvessel perfusion techniques are now well established in laboratories studying vascular permeability and could be adapted to allow dye perfusion of single capillaries or post-capillary venules in punch skin biopsies taken from dengue patients – dye perfusion must be carried out rapidly before dehydration occurs but then specimens can be fixed appropriately for EM studies. Immunohistochemical techniques are also being developed to visualise the glycocalyx in biopsy specimens, using fluorescently labelled monoclonal antibodies to constituents of the layer, such as heparan sulphate and hyaluronan binding protein.

5 Summary

In summary it is apparent that the spectrum of plasma leakage associated with dengue is broad and as techniques for identification of minor degrees of leakage become more sophisticated this is likely to become more apparent. It appears to be present in mild as well as severe disease, primary as well as secondary infections and starts earlier and persists for longer than previously understood. It is possible that all dengue-infected individuals experience some degree of vascular leak albeit at clinically undetectable levels in the majority of cases.

If we are to establish what happens at the microvascular level it is important to consider the endothelial barrier in its entirety rather than to focus only on changes that may occur at the cellular level in response to inflammatory mediators. Careful observational studies of human infection, together with an integrated approach utilising the knowledge, expertise and resources developed by physiologists, endothelial biologists and basic scientists to interrogate microvascular dysfunction in other situations, together with the experience of the dengue scientific community, may yet prove to be rewarding.

References

Cohen SN, Halstead SB (1966) Shock associated with dengue infection. I. Clinical and physiologic manifestations of dengue hemorrhagic fever in Thailand, 1964. J Pediatr 68(3):448–456

Nimmannitya S, Halstead SB, Cohen SN, Margiotta MR (1969) Dengue and chikungunya virus infection in man in Thailand, 1962–1964. I. Observations on hospitalized patients with hemorrhagic fever. Am J Trop Med Hyg 18(6):954–971

Shresta S, Sharar KL, Prigozhin DM, Beatty PR, Harris E (2006) Murine model for dengue virus-induced lethal disease with increased vascular permeability. J Virol 80(20):10208–10217

Bunyaratvej A, Butthep P, Yoksan S, Bhamarapravati N (1997) Dengue viruses induce cell proliferation and morphological changes of endothelial cells. Southeast Asian J Trop Med Public Health 28(Suppl 3):32–37

Avirutnan P, Malasit P, Seliger B, Bhakdi S, Husmann M (1998) Dengue virus infection of human endothelial cells leads to chemokine production, complement activation and apoptosis. J Immunol 161(11):6338–6346

Bonner SM, O'Sullivan MA (1998) Endothelial cell monolayers as a model system to investigate dengue shock syndrome. J Virol Methods 71(2):159–167

Lin CF, Lei HY, Shiau AL et al (2003) Antibodies from dengue patient sera cross-react with endothelial cells and induce damage. J Med Virol 69(1):82–90

Potter DR, Damiano ER (2008) The hydrodynamically relevant endothelial cell glycocalyx observed in vivo is absent in vitro. Circ Res 102(7):770–776

Michel CC, Curry FE (1999) Microvascular permeability. Physiol Rev 79(3):703–761

Weinbaum S, Tarbell JM, Damiano ER (2007) The structure and function of the endothelial glycocalyx layer. Annu Rev Biomed Eng 9:121–167

van den Berg BM, Vink H, Spaan JA (2003) The endothelial glycocalyx protects against myocardial edema. Circ Res 92(6):592–594

Perrin RM, Harper SJ, Bates DO (2007) A role for the endothelial glycocalyx in regulating microvascular permeability in diabetes mellitus. Cell Biochem Biophys 49(2):65–72

Guyton AC, Hall JE (2000) The microcirculation and the lymphatic system: capillary fluid exchange, interstitial fluid and lymph flow. In: Guyton AC (ed) Textbook of Medical Physiology, 10th edn. Saunders, WB, pp 162–173

Suwanik R, Tuchinda P, Tuchinda S et al (1967) Plasma volume and other third space studies in Thai haemorrhagic fever. Journal of the Medical Association of Thailand 50:48–66

Setiawan MW, Samsi TK, Pool TN, Sugianto D, Wulur H (1995) Gallbladder wall thickening in dengue hemorrhagic fever: an ultrasonographic study. J Clin Ultrasound 23(6):357–362

Thulkar S, Sharma S, Srivastava DN, Sharma SK, Berry M, Pandey RM (2000) Sonographic findings in grade III dengue hemorrhagic fever in adults. J Clin Ultrasound 28(1):34–37

Wu KL, Changchien CS, Kuo CH et al (2004) Early abdominal sonographic findings in patients with dengue fever. J Clin Ultrasound 32(8):386–388

Venkata Sai PM, Dev B, Krishnan R (2005) Role of ultrasound in dengue fever. Br J Radiol 78 (929):416–418

Colbert JA, Gordon A, Roxelin R et al (2007) Ultrasound measurement of gallbladder wall thickening as a diagnostic test and prognostic indicator for severe dengue in pediatric patients. Pediatr Infect Dis J 26(9):850–852

Srikiatkhachorn A, Krautrachue A, Ratanaprakarn W et al (2007) Natural history of plasma leakage in dengue hemorrhagic fever: a serial ultrasonographic study. Pediatr Infect Dis J 26(4):283–290 discussion 91–92

Statler J, Mammen M, Lyons A, Sun W (2008) Sonographic findings of healthy volunteers infected with dengue virus. J Clin Ultrasound 36(7):413–417

Gamble J, Gartside IB, Christ F (1993) A reassessment of mercury in silastic strain gauge plethysmography for microvascular permeability assessment in man. J Physiol 464:407–422

Starling EH (1896) On the absorption of fluids from the connective tissue spaces. J Physiol 19:312–326

Michel CC (1997) Starling: the formulation of his hypothesis of microvascular fluid exchange and its significance after 100 years. Exp Physiol 82(1):1–30

Gamble J, Bethell D, Day NP et al (2000) Age-related changes in microvascular permeability: a significant factor in the susceptibility of children to shock? Clin Sci (Lond) 98(2):211–216

Bethell DB, Gamble J, Pham PL et al (2001) Noninvasive measurement of microvascular leakage in patients with dengue hemorrhagic fever. Clin Infect Dis 32(2):243–253

Sahaphong S, Riengrojpitak S, Bhamarapravati N, Chirachariyavej T (1980) Electron microscopic study of the vascular endothelial cell in dengue hemorrhagic fever. Southeast Asian J Trop Med Public Health 11(2):194–204

Jessie K, Fong MY, Devi S, Lam SK, Wong KT (2004) Localization of dengue virus in naturally infected human tissues, by immunohistochemistry and in situ hybridization. J Infect Dis 189(8):1411–1418

Balsitis SJ, Coloma J, Castro G et al (2009) Tropism of dengue virus in mice and humans defined by viral nonstructural protein 3-specific immunostaining. Am J Trop Med Hyg 80(3):416–424

de Araujo JM, Schatzmayr HG, de Filippis AM et al (2009) A retrospective survey of dengue virus infection in fatal cases from an epidemic in Brazil. J Virol Methods 155(1):34–38

Adamson RH, Michel CC (1993) Pathways through the intercellular clefts of frog mesenteric capillaries. J Physiol 466:303–327

Majno G, Palade GE (1961) Studies on inflammation. 1. The effect of histamine and serotonin on vascular permeability: an electron microscopic study. J Biophys Biochem Cytol 11:571–605

Michel CC, Neal CR (1999) Openings through endothelial cells associated with increased microvascular permeability. Microcirculation 6(1):45–54

Schaeffer RC Jr, Gong F, Bitrick MS Jr, Smith TL (1993) Thrombin and bradykinin initiate discrete endothelial solute permeability mechanisms. Am J Physiol 264(6 Pt 2):H1798–H1809

Michel CC, Kendall S (1997) Differing effects of histamine and serotonin on microvascular permeability in anaesthetized rats. J Physiol 501(Pt 3):657–662

Myers BD, Guasch A (1993) Selectivity of the glomerular filtration barrier in healthy and nephrotic humans. Am J Nephrol 13(5):311–317

Blouch K, Deen WM, Fauvel JP, Bialek J, Derby G, Myers BD (1997) Molecular configuration and glomerular size selectivity in healthy and nephrotic humans. Am J Physiol 273(3 Pt 2): F430–F437

Oragui EE, Nadel S, Kyd P, Levin M (2000) Increased excretion of urinary glycosaminoglycans in meningococcal septicemia and their relationship to proteinuria. Crit Care Med 28(8): 3002–3008

Garcia S, Morales R, Hunter RF (1995) Dengue fever with thrombocytopenia: studies towards defining vulnerability of bleeding. Bol Asoc Med P R 87(1–2):2–7

Wills BA, Oragui EE, Dung NM et al (2004) Size and charge characteristics of the protein leak in dengue shock syndrome. J Infect Dis 190(4):810–818

Vink H, Duling BR (2000) Capillary endothelial surface layer selectively reduces plasma solute distribution volume. Am J Physiol Heart Circ Physiol 278(1):H285–H289

Michel CC, Curry FR (2009) Glycocalyx volume: a critical review of tracer dilution methods for its measurement. Microcirculation 16(3):213–219

Markers of Dengue Disease Severity

Anon Srikiatkhachorn and Sharone Green

Contents

Abstract Infection with one of the four serotypes of dengue virus (DENV) causes a wide spectrum of clinical disease ranging from asymptomatic infection, undifferentiated fever, dengue fever (DF) to dengue hemorrhagic fever (DHF). DHF occurs in a minority of patients and is characterized by bleeding and plasma leakage which may lead to shock. There are currently no reliable clinical or laboratory indicators that accurately predict the development of DHF. Human studies have shown that high viral load and intense activation of the immune system are associated with DHF. Recently, endothelial cells and factors regulating vascular permeability have been demonstrated to play a role. In the absence of animal models that closely mimic DHF, human studies are essential in identifying predictors of severe illness.

A. Srikiatkhachorn (✉) and S. Green

Center for Infectious Disease and Vaccine Research, University of Massachusetts Medical School, 55 Lake Avenue, North, Worcester, 01655, USA
e-mail: anon.srikiatkhachorn@umassmed.edu

A.L. Rothman (ed.), *Dengue Virus*, Current Topics in Microbiology and Immunology 338,   67
DOI 10.1007/978-3-642-02215-9_6, © Springer-Verlag Berlin Heidelberg 2010

Well planned prospective studies with samples collected at different time points of the illness in well characterized patients are crucial for this effort. Ideally, clinical and laboratory predictive tools should be suitable for resource poor countries where dengue is endemic.

1 Introduction

1.1 Spectrum of Clinical Manifestations in Dengue

Infection with one of the four serotypes of dengue virus (DENV) may cause a wide spectrum of clinical disease (Nimmannitya 1993). The majority of DENV infections in children are clinically inapparent, although some may develop undifferentiated fever. Primary infections in older children and adults are more likely to cause dengue fever (DF), a febrile illness accompanied by nonspecific symptoms including headache, retroorbital pain, myalgia and hemorrhagic manifestations. A minority of patients develop dengue hemorrhagic fever (DHF), the more severe form of dengue disease, whose hallmark is plasma leakage leading to intravascular volume loss and circulatory insufficiency. Bleeding is common in both DF and DHF but more severe bleeding, particularly bleeding from the gastrointestinal tract is found more frequently in DHF than in DF. (Nimmannitya 1993). There have been reports of other less frequent but severe clinical manifestations including hepatic failure and encephalopathy, which may be consequences of shock or may represent distinct clinical manifestations of dengue illness. (Cam et al. 2001; Kalayanarooj and Nimmannitya 2003; Pancharoen et al. 2002; Huerre et al. 2001)

1.2 Sequence of Clinical Events in DENV Infection

An appreciation of the clinical progression of dengue infection is a prerequisite to the understanding of the underlying biological processes leading to severe manifestations of dengue disease. Patients with DF or DHF typically present with a history of abrupt onset of high, persistent fever (Nimmannitya 1993). Other common clinical manifestations during the febrile phase of the illness include myalgia, nausea, vomiting and abdominal pain. During this period patients display varied degrees of hemorrhagic tendency ranging from petechiae to epistaxis or gastrointestinal bleeding. Significant dehydration may develop at this stage of illness, which may require intravenous fluid treatment. The febrile period lasts 2–7 days. Around the time of defervescence patients may develop localized plasma leakage that manifests as accumulation of fluid in the pleural or abdominal cavities and hemoconcentration. The presence of plasma leakage, the key feature in differentiating DHF from DF, lasts approximately 48 h and is followed by spontaneous and rapid resolution. Clinically significant bleeding may also occur during this period, particularly in children with untreated circulatory insufficiency. In addition, hepatic failure and encephalopathy may develop secondary to prolonged shock. Mortality is usually secondary to a delay in the recognition and treatment of plasma leakage.

1.3   Interpreting Studies of Markers of Severe Dengue Disease

Due to the lack of an animal model that can reproduce the disease seen in humans, clinical studies have been instrumental to the understanding of the pathogenesis of this disease. Numerous studies have focused on identifying biological parameters in clinical samples that differ between severe and nonsevere cases. Studies of biological markers may provide early predictors that can identify cases that will eventually develop DHF. This is critical, since currently there are no routine clinical or laboratory parameters that can predict the severity of individual cases, necessitating close monitoring of all dengue patients. Furthermore, such studies may help identify novel approaches for therapy of plasma leakage and hemorrhage in severe dengue cases.

There are several cautions in interpreting these studies. In most studies, only a single sample was collected from each patient at various time points during the clinical course. The heterogeneity in the timing of sample collection may obfuscate differences that exist between patients with different disease severity. In addition, since severe manifestations of dengue disease mostly develop around the time of defervescence, changes in biological markers should be analyzed in the context of time of defervescence rather than day after onset. Uniformity of clinical diagnosis and grading of dengue cases is a prerequisite for the analysis of the changes in biological markers in dengue patients (Table 1). Differences in the processing of samples, such as plasma versus serum, may have a significant impact on the levels of markers of interest, particularly those that may be released or activated during coagulation.

2   Clinical Laboratory Markers of Severe Disease

2.1   Platelets

Marked thrombocytopenia (platelet count < 100,000 cells per cu mm) was recognized in the original description of DHF and is one of the current criteria for the diagnosis of DHF. (Nimmannitya 1993; Zarco et al. 1957) Although thrombocytopenia is more common in DHF than DF, a significant fraction of DF patients also develop thrombocytopenia. Thrombocytopenia is not an early indicator for DHF as the platelet counts during the early febrile phase of DF and DHF are not

Table 1 Changes in biological markers in dengue infections

Biological markers Cytokines	Comparison	Timing of peak response	Comments
IFN-α	DHF, DF > healthy control (Kurane et al. (1993))	Febrile phase	
IFN-γ	Severe > mild[a] (Bozza et al. (2008)) Not different between DF and DHF (Braga et al. (2001); Kurane et al. (1991)) DF > healthy control (Azeredo et al. (2001))	Not specified (Bozza et al. (2008)) Febile phase (Braga et al. (2001); Kurane et al. (1991)) Acute illness, timing in reference to fever unspecified. (Azeredo et al. (2001))	
IL-2	Not different between DF and DHF (Kurane et al. (1991))	Early febile phase (Kurane et al. (1991))	
IL-6	DHF non survivor > DHF survivor DHF > DF (Chen et al. (2006); Suharti et al. (2003)) Severe > mild[a] (Bozza et al. (2008)) DF=DHF=other febrile illnesses (Green et al. (1999c))	Acute illness, timing in reference to fever unspecified. (Chen et al. (2006); Subarti et al. (2003)) Not specified (Bozza et al. (2008)) Acute illness (Green et al. (1999c))	
IL-8	DSS > healthy control (Avirutnan et al. (1998)) DHF grade 3,4 > DHF grade 1,2 and DF (Raghupathy et al. (1998))	Not specified (Avirutnan et al. (1998)) Acute illness, timing in reference to fever unspecified. (Raghupathy et al. (1998))	Also present in pleural fluid (Avirutnan et al. (1998))
IL-10	DHF > DF (Chen et al. (2006)) DHF, DF > healthy control (Azeredo et al. (2001)) DHF > DF (Green et al. (1999b); Libraty et al. (2002a))	Acute illness (Chen et al. (2006)) Acute illness, timing in reference to fever unspecified. Azeredo et al. (2001) Acute febrile and defervescence phase Green et al. (1999b); Libraty et al. (2002a)	Levels correlate with the size of pleural effusions. (Green et al. (1999b))
MIF (Migration inhibitory factor)	DHF non survivor > DHF survivor DHF > DF. (Chen et al. (2006))	Acute illness, (Chen et al. (2006)) timing in reference to fever unspecified.	
TNF-α	DHF, DF > healthy control (Azeredo et al. (2001)) DHF, DF with hemorrhage > DF (Braga et al. (2001))	Acute illness, timing in reference to fever unspecified. (Azeredo et al. (2001))	

(continued)

Table 1 (continued)

Biological markers Cytokines	Comparison	Timing of peak response	Comments
	DHF grade 3,4 > DHF grade 1,2 (Hober et al. (1993))	Acute illness, timing in reference to fever unspecified. (Braga et al. (2001)) Acute illness, timing in reference to fever unspecified. (Hober et al. (1993))	
CXCL9, CXCL10, CXCL11	DHF > DF (Dejnirattisai et al. (2008))	Febrile phase (Dejnirattisai et al. (2008))	
MCP-1	DSS > healthy control (Avirutnan et al. (1998))	Not specified (Avirutnan et al. (1998))	Also present in pleural fluid (Avirutnan et al. (1998))
VEGF	DHF > DF (Srikiatkhachorn et al. (2007); Tseng et al. (2005))	Defervescence (Srikiatkhachorn et al. (2007)) Acute illness, timing in reference to fever unspecified. (Tseng et al. (2005))	
Coagulation and endothelial markers			
Von Willebrand factor	DHF > DF (Sosothikul et al. (2007))	Febrile and defervescence phase (Sosothikul et al. (2007))	
Tissue factor	DHF > DF (Sosothikul et al. (2007))	Febrile and defervescence phase (Sosothikul et al. (2007))	
Plasminogen activator inhibitor	DHF > DF (Sosothikul et al. (2007))	Febrile and defervescence phase (Sosothikul et al. (2007))	
Soluble thrombo-modulin	DHF grade 3,4 > DHF grade 1,2 and DF (Butthep et al. (2006)) DHF > DF (Sosothikul et al. (2007))	Febrile phase (Butthep et al. (2006)) Febrile phase (Sosothikul et al. (2007))	
sVEGFR2	DHF < DF (Srikiatkhachorn et al. (2007))	Febrile and defervescence phase (Srikiatkhachorn et al. (2007))	
sICAM-1	Dengue > healthy control Not different between DF and DHF (Valero et al. (2008); Cardier et al. (2006))	Acute illness, timing in reference to fever unspecified (Valero et al. (2008); Cardier et al. (2006))	
sVCAM	DHF > healthy control (Cardier et al. (2006)) DHF grade 3,4 > DHF grade 1,2 and DF (Koraka et al. (2004))	Acute illness, timing in reference to fever unspecified. (Cardier et al. (2006))	

(*continued*)

Table 1 (continued)

Biological markers	Comparison	Timing of peak response	Comments
Cytokines			
		Acute illness, timing in reference to fever unspecified (Koraka et al. (2004))	
Other Soluble receptors			
sCD4	DHF>DF (Kurane et al. (1991)) DF=DHF=other febrile illnesses (Green et al. (1999c))	Acute illness, timing in reference to fever unspecified. (Kurane et al. (1991))	
sCD8	DHF>DF (Kurane et al. (1991)) DHF>DF, DHF, DF> other febrile illnesses (Green et al. (1999c))	Acute illness, timing in reference to fever unspecified. (Kurane et al. (1991)) Defervescence (Green et al. (1999c))	
sTNFRII (p75)	DHF=DF (Braga et al. (2001)): DSS>DHF (Bethell et al. 1998)) DF>healthy control (Azeredo et al. (2001)) DHF>DF, DHF, dengue>other febrile illnesses (Green et al. (1999c))	Not specified (Braga et al. (2001); Bethell et al. (1998)) Acute illness, timing in reference to fever unspecified. (Azeredo et al. (2001)) Febrile and defervescence phase (Green et al. (1999c))	sTNFRII levels correlate with the size of pleural effusions.
IL-1Ra	DHF non survivor>DHF survivor (Suharti et al. (2003))	Febrile phase (Suharti et al. (2003))	
sIL-2R	Dengue>healthy control DHF grade 3>DHF grade,1, 2 (Valero et al. (2008)) DHF>DF>healthy control (Kurane et al. (1991)) DHF>DF (Libraty et al. (2002a); Green et al. (1999c))	Acute illness, timing in reference to fever unspecified (Valero et al. (2008)) Febrile phase (Kurane et al. (1991)) Febrile and defervescence phase Libraty et al. (2002a)	Correlates with elevated liver enzyme (Libraty et al. (2002a))

DHF (dengue hemorrhagic fever) *DF* (dengue fever), *DSS* (dengue shock syndrome)
[a]The definition of severe is the presence of any of the followings findings: (1) thrombocytopenia ($<$50,000 per cu mm), (2) hypotension, (3) evidence of plasma leakage or hemoconcentration

significantly different. Platelet counts progressively decline during the febrile phase in both DF and DHF cases, reaching their lowest point at the time of defervescence in both DF and DHF and coinciding with plasma leakage in patients with DHF. As such, platelet counts serve as a monitoring tool for disease progression rather than an early indicator for severe disease. In our analysis that compared dengue cases requiring fluid treatment to those that did not, we found that a platelet count of 60,000 cells per cu mm served as a better cut-off in identifying more severe cases (unpublished observation). Platelet counts are rarely low enough to cause spontaneous hemorrhage in DHF patients but may contribute to the hemorrhagic tendency in cases complicated with plasma leakage and shock. Platelet counts showed an inverse correlation with the size of pleural effusions in DHF patients, suggesting that platelets can serve as a marker for the extent of plasma leakage (Nimmannitya 1993).

Immune complexes containing dengue antigen have been reported on platelet surfaces and may be one mechanism underlying the increased platelet destruction. (Wang et al. 1995; Mitrakul et al. 1977; Phanichyakarn et al. 1977) Platelet-associated antibodies have been reported to be linked to decreased platelet counts and clinical severity in acute secondary DENV infections. (Saito et al. 2004) The mechanisms of the association were not delineated in these studies but the antibodies eluted from platelets had anti-DENV activity. Subsequent studies have shown that antibodies to NS1 antigen cross-react with platelets leading to platelet activation and opsonization in vitro. (Sun et al. 2007) In vivo, administration of anti-NS1 antibodies into warfarin-treated mice led to a decrease in platelet counts and appeared to induce some degree of activation of the coagulation pathway. Interestingly, in vivo treatment with this antibody did not cause hemorrhage in experimental animals but instead caused plasma leakage in the lungs, a pattern of leakage distinct from the plasma leakage into pleural or peritoneal spaces found in DHF (Sun et al. 2007). In view of the rapid recovery of DHF patients and the lack of any autoimmune diseases afterward, the relevance of these antiplatelet antibodies in the pathogenesis of dengue is unclear.

2.2 Liver Enzymes

Hepatomegaly and liver tenderness are common findings in DF and DHF. Elevated liver transaminases (AST and ALT) occur early in the course of the disease and have been reported to be more pronounced in DHF (Kalayanarooj and Nimmannitya 2003; Pancharoen et al. 2002; Kalayanarooj et al. 1997). AST and ALT may serve as early markers of severe dengue disease; however, laboratory value cut-offs that will optimally differentiate DF from DHF cases have not yet been defined.

Liver pathology of autopsy cases has demonstrated mild infiltration of monocytes and lymphocytes and varying degrees of necrosis (Bhamarapravati et al. 1967). DENV antigen was localized in the cytoplasm of Kuppfer cells and endothelium but not in hepatocytes in one study (Jessie et al. 2004). A study using in situ PCR and immunohistochemistry demonstrated DENV genome in the majority of hepatocytes and Kuppfer cells in some but not all fatal dengue cases (Huerre et al. 2001). In vitro, DENV-specific CD4+ T cells have been demonstrated to mediate bystander cytotoxicity to hepatocytes via the Fas pathway (Gagnon et al. 1999).

In vivo, CD8+ T cells appeared to be involved in liver pathology in experimental models of DENV infection in mice (Chen et al. 2004). Taken together, these findings suggested that liver pathology may be mediated by both virus as well as host mediated immune responses against the virus.

3 Biological Markers in the Early Febrile Phase of the Disease

3.1 Dengue Virus and Dengue Viral Products

Studies using virologic and molecular techniques to measure plasma viral load have demonstrated that patients with DHF had higher viral loads than patients with DF (Libraty et al. 2002a; Vaughn et al. 2000). The peak viremia appears to occur early in the course of the disease, followed by a precipitous drop of plasma viral titers at defervescence. The plasma levels of a viral product, soluble NS1 antigen, were also higher in DHF cases than in DF cases within the first 72 h of onset of fever. (Libraty et al. 2002b) These results demonstrated that higher viral load is a risk factor for severe disease.

DENV NS1 antigen has been demonstrated in vitro to activate complement by itself and this activity is enhanced by the presence of DENV-specific antibody. (Avirutnan et al. 2006) The molecular mechanism of the complement fixing activity of NS1 alone is currently unknown. The role of the complement fixing activity of NS1 in plasma leakage has been questioned due to the incongruity of the timing of peak NS1 levels, which occurs during the febrile phase and the timing of plasma leakage, which occurs days later. Nevertheless, levels of NS1 antigen in biological samples may be a useful early indicator of severe dengue disease.

3.2 Cytokines of the Innate Immune System

Infection of cellular targets such as monocytes/macrophages, dendritic cells and possibly endothelial cells with DENV induces production of several cytokines including TNF-α, IL-1, IL-6, IL-10 and chemokines, depending on the infected cell type (Ho et al. 2001; Bosch et al. 2002; Huang et al. 2000; Libraty et al. 2001; Avirutnan et al. 1998; Kurane et al. 1992). Circulating levels of some of these cytokines have been measured in dengue patients with conflicting results. This is likely due to several factors, including differences in the timing of sample collection, detection methods and the accuracy in assigning disease severity. Cytokines with proinflammatory and vascular permeability-enhancing activity such as TNF-α and IL-8 have been reported to be elevated in DHF but the role of these cytokines in plasma leakage in DHF remains unclear (Avirutnan et al. 1998; Laur et al. 1998; Wang et al. 2007; Braga et al. 2001). The relative lack of tissue inflammation in

DHF appears to argue against these cytokines as major agents causing plasma leakage in this condition. Regardless of their contribution to plasma leakage, elevated levels of these cytokines may serve as early markers for severe disease. More information is needed related to the timing of these cytokine elevations to assess their usefulness as early markers of severity.

Studies of cytokine levels in serial plasma samples from DENV-infected patients have demonstrated characteristic patterns of cytokine responses during the course of DENV infection (Fig. 1). The cytokines that were induced during the early febrile phase were the alpha interferons, reflecting activation of the innate immune system. (Libraty et al. 2002a; Navarro-Sanchez et al. 2005; Kurane and Ennis 1988) In a nonhuman primate model, circulating plasmacytoid dendritic cells, a potent interferon-α producer, increased in number soon after viral inoculation and persisted during the viremic phase (Pichyangkul et al. 2003). In human DHF cases, circulating plasmacytoid dendritic cells and their precursors showed a blunted response compared to DF patients but plasma interferon-α levels did not differ (Pichyangkul et al. 2003). IFN-α production by PBMC in response to oligonucleotide stimulation appeared to be somewhat impaired in dengue patients when compared to normal PBMC. However, no difference in IFN-α production was observed when PBMC from DHF and DF patients were stimulated with oligonucleotide (Pichyangkul et al. 2003). The effectiveness of IFN-α in controlling DENV replication in a primate model supports the role of this cytokine as an important first

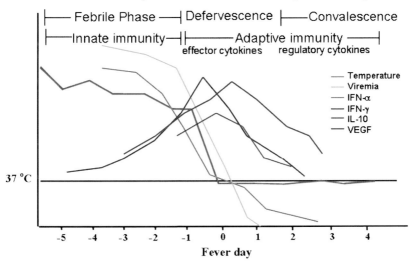

Fig. 1 Temperature, virological and biological parameters during the course of dengue virus infection. Fever day 0 indicates the day of defervescence; fever days −1, −2, etc., indicate days prior to defervescence and fever days +1, +2, etc., indicate days after defervescence, respectively. Plasma leakage in DHF usually occurs around the time of defervescence, accompanied by a rapid decline in viremia

line of defense (Ajariyakhajorn et al. 2005). A defect in the maturation or recruit-ment of plasmacytoid dendritic cells may contribute to more severe disease and may serve as an early indicator of severe disease.

NK cells play an important role in the innate immune system against virus infection. Flow cytometric evidence of early activation of NK cells has been demonstrated in dengue cases and the activation appeared to be more pronounced in DHF cases in comparison to DF cases (Green et al. 1999a; Azeredo et al. 2006; Homchampa et al. 1988). Elevated levels of IFN-γ and TNF-α , both of which can be secreted by NK cells, have been reported in DHF cases (Braga et al. 2001; Avila-Aguero et al. 2004; Kurane et al. 1991). However, the peak levels of these cytokines, which occur at the time of declining viremia, suggest that these cytokines are produced from cells of the adaptive immune system as well. It is likely that the elevated levels of IFN-γ and TNF-α reflect an enhanced activation of both the adaptive immune system and the subsequent amplified response of NK cells.

4   Biological Markers During Late Febrile Phase and the Defervescence Phase of the Disease

4.1   Cytokines of the Adaptive Immune Response

A number of studies have demonstrated a more intense cellular immune response in DHF compared to DF. During a secondary infection with a distinct DENV serotype, serotype-cross-reactive T cells are activated, resulting in elevated levels of IL-2, IFN-γ and TNF-α, which have been reported to be more pronounced in DHF compared to DF in some studies. (Braga et al. 2001; Avila-Aguero et al. 2004; Kurane et al. 1991; Bozza et al. 2008) Other studies, however, have failed to demonstrate these differences (Chen et al. 2006). The increase in levels of these cytokines occurs after the peak levels of IFN-α and viremia but slightly before or at the time of defervescence and plasma leakage (Libraty et al. 2002a). Some of these cytokines, particularly TNF-α, have a direct permeability-enhancing effect and a coagulation-activating effect. Th2-type cytokines, including IL-4 and IL-13, have also been reported to be elevated in DHF but the roles of these cytokines in the pathogenesis of DHF are unclear (Bozza et al. 2008; Mustafa et al. 2001). Elevated levels of IL-10 have been demonstrated in DHF patients compared to DF patients, possibly reflecting an enhanced regulatory response to the more intense immune activation in DHF (Green et al. 1999b). In addition to T cells, macrophages infected with DENV-antibody immune complexes may be a source of IL-10. IL-10 levels measured two days prior to defervescence correlated with the size of the pleural effusion detected one day following defervescence, suggesting that IL-10 might be a candidate marker of severe disease. However, IL-10 levels appeared to peak after defervescence in another study (Libraty et al. 2002a). This may reflect the distinct sources of IL-10 including infected macrophage during the early phase of the disease and regulatory T cells in the later phase of the illness.

4.2   Alteration in Soluble Receptors as Markers for Severe Disease

Soluble forms of cellular receptors for cytokines and other ligands can be generated by various mechanisms including differential splicing or enzymatic cleavage of surface molecules. Earlier studies demonstrated elevated levels of soluble CD4 and CD8 in the plasma of DHF patients (Kurane et al. 1991). Elevated levels of soluble forms of TNF-α receptor in DHF have been reported in some but not all studies (Libraty et al. 2002a; Wang et al. 2007; Braga et al. 2001; Green et al. 1999c). These levels have been shown to correlate with the size of pleural effusions, an indicator of plasma leakage in vivo (Green et al. 1999c). Elevated levels of soluble IL-2 receptor have been reported to be correlated with the increase in liver enzymes in one study but not another. (Libraty et al. 2002a; Valero et al. 2008) Serum levels of IL-1Ra were reported to be higher in fatal dengue cases compared to the levels found in survivors. (Suharti et al. 2003) A recent study has demonstrated elevated serum levels of soluble ST2, a soluble form of the interleukin-1 receptor-like 1 protein, in dengue cases compared to patients with other febrile illnesses. (Becerra et al. 2008) The levels correlated with the decline in white blood cells and the increase in serum transaminase levels. (Valero et al. 2008) Whether these correlations reflect the roles of these receptors in the pathogenesis of DHF and the mechanisms by which these molecules may modulate disease severity are unknown.

We and others have reported increased levels of a potent permeability enhancing cytokine, vascular endothelial growth factor (VEGF), in DHF patients during plasma leakage (Srikiatkhachorn et al. 2007; Tseng et al. 2005). This increase was associated with a progressive decline of the circulating soluble form of VEGFR-2 receptor in the plasma of DHF patients. The decline in soluble VEGFR-2 was accompanied by a decline in soluble VEGFR-2 – VEGF complexes and a rise in the biologically active, free form of VEGF. The degree of decline in plasma soluble VEGFR-2 correlated with the extent of plasma leakage in vivo. In addition to its permeability-enhancing property, VEGF also up-regulates the production of tissue factor, leading to activation of the coagulation system (Shen et al. 2001). These properties make VEGF an attractive candidate cytokine that may play a role in the two main features in DHF, plasma leakage and coagulopathy. Infection of endothelial cells in vitro resulted in a suppression of soluble VEGFR2 production and a concomitant increase in surface expression of this molecule (Srikiatkhachorn et al. 2007). In vivo, the degree of decline in soluble VEGFR2 correlated with plasma viral load. This represents a novel mechanism by which DENV may regulate vascular permeability, by altering membrane and soluble forms of receptors for vasoactive cytokines.

4.3 Markers of Endothelial Cell Activation and Injury

Endothelial cells play the key role in the final stages of the pathogenesis of DHF. Activation of endothelial cells leads to changes in vascular permeability and release of factors that activate the coagulation pathway. In vitro, DENV-infected endothelial cells have been shown to produce several chemokines including IL-8 and RANTES (Bosch et al. 2002; Avirutnan et al. 1998). Very high levels of IL-8 have been reported in plasma and pleural fluids of patients with DHF grade IV (Avirutnan et al. 1998; Raghupathy et al. 1998). Infection of endothelial cells also led to the activation of complement and cellular apoptosis (Avirutnan et al. 1998). The relevance of these findings to the pathogenesis of DHF is unclear for several reasons. Pleural and ascitic fluids in DHF do not contain neutrophils or other inflammatory cells. In addition, the evidence of endothelial cell apoptosis in vivo is very limited (Limonta et al. 2007). Lastly, the rapid recovery of plasma leakage in DHF argues against structural damage of blood vessels as the underlying mechanism of plasma leakage.

Endothelial cell functions may be affected by cytokines released from other cell types that are infected with DENV. Studies have shown that supernatants from DENV-infected monocytes caused upregulation of ICAM-1 expression by endothelial cells, which is likely mediated by TNF-α (Anderson et al. 1997). Elevated levels of soluble endothelial surface molecules such as ICAM-1 and VCAM have been reported in DHF patients (Koraka et al. 2004; Cardier et al. 2006). In addition, the frequencies of circulating endothelial cells have been reported to be increased in DHF patients (Cardier et al. 2006; Butthep et al. 2006). Perturbation of the coagulation system is closely associated with endothelial cell activation. Increased levels of soluble thrombomodulin, von Willebrand factor antigen, tissue factor and plasminogen activator have been reported in dengue patients during the acute phase and appear to be associated with disease severity (Butthep et al. 2006; Sosothikul et al. 2007). In contrast, the levels of ADAMTS-13, a von Willebrand factor-cleaving metalloprotease, were reported to be lower in DHF patients compared to DF patients (Sosothikul et al. 2007). Taken together, these findings suggest activation of endothelial cells in vivo. Analysis of endothelial cell activation markers may therefore be helpful in identifying severe dengue cases and provide some insight into the pathogenesis of this disease. However, kinetic studies are needed to delineate the time course of the levels of markers of endothelial activation and to evaluate whether these markers can be used as a predictor of severe disease early in the course of the illness.

5 Conclusions and Future Directions

A recent increase in dengue research underscores the importance of this disease as a global health threat. One of the challenges in the care of dengue patients is the inability in the early stage of illness to reliably identify patients that will subsequently develop severe disease. Biological markers that can predict or serve as correlates for disease severity are urgently needed. In the absence of animal models that can recapitulate human DHF, studies in dengue patients remain indispensable. Difficulties in interpreting and assimilating information from human studies lie in the heterogeneity in study designs, the timing and the methods of sample collection and processing and the accuracy in assigning disease severity. In light of these problems, prospective studies with well-characterized patients with different degrees of severity are crucial in the effort to better understand the underlying mechanisms of this disease. In an ideal setting, samples should be prospectively collected from clinically well-defined cohorts at various time points before, during and after defervescence and properly processed for the measurement of the marker(s) of interest. Such studies will provide much needed information on the kinetics of changes in biological markers and how these may relate to disease severity.

The limitation in the volume of samples obtainable from dengue patients, who are mostly children, necessitates a carefully planned analysis that will yield maximum useful information. Newly developed techniques capable of measuring multiple analytes and multiple gene transcripts should become important for research in this area. Integration of clinical and laboratory parameters such as viral load, cytokines and cytokine receptors and indicators of cellular activation and transcription will be crucial in developing a predictive algorithm capable of identifying severe cases early in the course of the illness. Importantly, such algorithms must be applicable in settings with limited resources such as dengue-endemic countries where these tools are most needed.

References

Ajariyakhajorn C, Mammen MP Jr, Endy TP, Gettayacamin M, Nisalak A, Nimmannitya S, Libraty DH (2005) Randomized, placebo-controlled trial of nonpegylated and pegylated forms of recombinant human alpha interferon 2a for suppression of dengue virus viremia in rhesus monkeys. Antimicrob Agents Chemother 49:4508–4514

Anderson R, Wang S, Osiowy C, Issekutz AC (1997) Activation of endothelial cells via antibody-enhanced dengue virus infection of peripheral blood monocytes. J Virol 71:4226–4232

Avila-Aguero ML, Avila-Aguero CR, Um SL, Soriano-Fallas A, Canas-Coto A, Yan SB (2004) Systemic host inflammatory and coagulation response in the Dengue virus primo-infection. Cytokine 27:173–179

Avirutnan P, Malasit P, Seliger B, Bhakdi S, Husmann M (1998) Dengue virus infection of human endothelial cells leads to chemokine production, complement activation, and apoptosis. J Immunol 161:6338–6346

Avirutnan P, Punyadee N, Noisakran S, Komoltri C, Thiemmeca S, Auethavornanan K, Jairungsri A, Kanlaya R, Tangthawornchaikul N, Puttikhunt C et al (2006) Vascular leakage in severe dengue virus infections: a potential role for the nonstructural viral protein NS1 and complement. J Infect Dis 193:1078–1088

Azeredo EL, Zagne SM, Santiago MA, Gouvea AS, Santana AA, Neves-Souza PC, Nogueira RM, Miagostovich MP, Kubelka CF (2001) Characterisation of lymphocyte response and cytokine patterns in patients with dengue fever. Immunobiology 204:494–507

Azeredo EL, De Oliveira-Pinto LM, Zagne SM, Cerqueira DI, Nogueira RM, Kubelka CF (2006) NK cells, displaying early activation, cytotoxicity and adhesion molecules, are associated with mild dengue disease. Clin Exp Immunol 143:345–356

Becerra A, Warke RV, de Bosch N, Rothman AL, Bosch I (2008) Elevated levels of soluble ST2 protein in dengue virus infected patients. Cytokine 41:114–120

Bethell DB, Flobbe K, Cao XT, Day NP, Pham TP, Buurman WA, Cardosa MJ, White NJ, Kwiatkowski D (1998) Pathophysiologic and prognostic role of cytokines in dengue hemorrhagic fever. J Infect Dis 177:778–782

Bhamarapravati N, Tuchinda P, Boonyapaknavik V (1967) Pathology of Thailand haemorrhagic fever: a study of 100 autopsy cases. Ann Trop Med Parasitol 61:500–510

Bosch I, Xhaja K, Estevez L, Raines G, Melichar H, Warke RV, Fournier MV, Ennis FA, Rothman AL (2002) Increased production of interleukin-8 in primary human monocytes and in human epithelial and endothelial cell lines after dengue virus challenge. J Virol 76:5588–5597

Bozza FA, Cruz OG, Zagne SM, Azeredo EL, Nogueira RM, Assis EF, Bozza PT, Kubelka CF (2008) Multiplex cytokine profile from dengue patients: MIP-1beta and IFN-gamma as predictive factors for severity. BMC Infect Dis 8:86

Braga EL, Moura P, Pinto LM, Ignacio SR, Oliveira MJ, Cordeiro MT, Kubelka CF (2001) Detection of circulant tumor necrosis factor-alpha, soluble tumor necrosis factor p75 and interferon-gamma in Brazilian patients with dengue fever and dengue hemorrhagic fever. Mem Inst Oswaldo Cruz 96:229–232

Butthep P, Chunhakan S, Tangnararatchakit K, Yoksan S, Pattanapanyasat K, Chuansumrit A (2006) Elevated soluble thrombomodulin in the febrile stage related to patients at risk for dengue shock syndrome. Pediatr Infect Dis J 25:894–897

Cam BV, Fonsmark L, Hue NB, Phuong NT, Poulsen A, Heegaard ED (2001) Prospective case-control study of encephalopathy in children with dengue hemorrhagic fever. Am J Trop Med Hyg 65:848–851

Cardier JE, Rivas B, Romano E, Rothman AL, Perez-Perez C, Ochoa M, Caceres AM, Cardier M, Guevara N, Giovannetti R (2006) Evidence of vascular damage in dengue disease: demonstration of high levels of soluble cell adhesion molecules and circulating endothelial cells. Endothelium 13:335–340

Chen HC, Lai SY, Sung JM, Lee SH, Lin YC, Wang WK, Chen YC, Kao CL, King CC, Wu-Hsieh BA (2004) Lymphocyte activation and hepatic cellular infiltration in immunocompetent mice infected by dengue virus. J Med Virol 73:419–431

Chen LC, Lei HY, Liu CC, Shiesh SC, Chen SH, Liu HS, Lin YS, Wang ST, Shyu HW, Yeh TM (2006) Correlation of serum levels of macrophage migration inhibitory factor with disease severity and clinical outcome in dengue patients. Am J Trop Med Hyg 74:142–147

Dejnirattisai W, Duangchinda T, Lin CL, Vasanawathana S, Jones M, Jacobs M, Malasit P, Xu XN, Screaton G, Mongkolsapaya J (2008) A complex interplay among virus, dendritic cells, T cells, and cytokines in dengue virus infections. J Immunol 181:5865–5874

Gagnon SJ, Ennis FA, Rothman AL (1999) Bystander target cell lysis and cytokine production by dengue virus-specific human CD4(+) cytotoxic T-lymphocyte clones. J Virol 73:3623–3629

Green S, Pichyangkul S, Vaughn DW, Kalayanarooj S, Nimmannitya S, Nisalak A, Kurane I, Rothman AL, Ennis FA (1999a) Early CD69 expression on peripheral blood lymphocytes from children with dengue hemorrhagic fever. J Infect Dis 180:1429–1435

Green S, Vaughn DW, Kalayanarooj S, Nimmannitya S, Suntayakorn S, Nisalak A, Rothman AL, Ennis FA (1999b) Elevated plasma interleukin-10 levels in acute dengue correlate with disease severity. J Med Virol 59:329–334

Green S, Vaughn DW, Kalayanarooj S, Nimmannitya S, Suntayakorn S, Nisalak A, Lew R, Innis BL, Kurane I, Rothman AL et al (1999c) Early immune activation in acute dengue illness is related to development of plasma leakage and disease severity. J Infect Dis 179:755–762

Ho LJ, Wang JJ, Shaio MF, Kao CL, Chang DM, Han SW, Lai JH (2001) Infection of human dendritic cells by dengue virus causes cell maturation and cytokine production. J Immunol 166:1499–1506

Hober D, Poli L, Roblin B, Gestas P, Chungue E, Granic G, Imbert P, Pecarere JL, Vergez-Pascal R, Wattre P et al (1993) Serum levels of tumor necrosis factor-alpha (TNF-alpha), interleukin-6 (IL-6), and interleukin-1 beta (IL-1 beta) in dengue-infected patients. Am J Trop Med Hyg 48:324–331

Homchampa P, Sarasombath S, Suvatte V, Vongskul M (1988) Natural killer cells in dengue hemorrhagic fever/dengue shock syndrome. Asian Pac J Allergy Immunol 6:95–102

Huang YH, Lei HY, Liu HS, Lin YS, Liu CC, Yeh TM (2000) Dengue virus infects human endothelial cells and induces IL-6 and IL-8 production. Am J Trop Med Hyg 63:71–75

Huerre MR, Lan NT, Marianneau P, Hue NB, Khun H, Hung NT, Khen NT, Drouet MT, Huong VT, Ha DQ et al (2001) Liver histopathology and biological correlates in five cases of fatal dengue fever in Vietnamese children. Virchows Arch 438:107–115

Jessie K, Fong MY, Devi S, Lam SK, Wong KT (2004) Localization of dengue virus in naturally infected human tissues, by immunohistochemistry and in situ hybridization. J Infect Dis 189:1411–1418

Kalayanarooj S, Nimmannitya S (2003) Clinical presentations of dengue hemorrhagic fever in infants compared to children. J Med Assoc Thai 86(Suppl 3):S673–680

Kalayanarooj S, Vaughn DW, Nimmannitya S, Green S, Suntayakorn S, Kunentrasai N, Viramitrachai W, Ratanachu-eke S, Kiatpolpoj S, Innis BL et al (1997) Early clinical and laboratory indicators of acute dengue illness. J Infect Dis 176:313–321

Koraka P, Murgue B, Deparis X, Van Gorp EC, Setiati TE, Osterhaus AD, Groen J (2004) Elevation of soluble VCAM-1 plasma levels in children with acute dengue virus infection of varying severity. J Med Virol 72:445–450

Kurane I, Ennis FA (1988) Production of interferon alpha by dengue virus-infected human monocytes. J Gen Virol 69(Pt 2):445–449

Kurane I, Innis BL, Nimmannitya S, Nisalak A, Meager A, Janus J, Ennis FA (1991) Activation of T lymphocytes in dengue virus infections. High levels of soluble interleukin 2 receptor, soluble CD4, soluble CD8, interleukin 2, and interferon-gamma in sera of children with dengue. J Clin Invest 88:1473–1480

Kurane I, Janus J, Ennis FA (1992) Dengue virus infection of human skin fibroblasts in vitro production of IFN-beta, IL-6 and GM-CSF. Arch Virol 124:21–30

Kurane I, Innis BL, Nimmannitya S, Nisalak A, Meager A, Ennis FA (1993) High levels of interferon alpha in the sera of children with dengue virus infection. Am J Trop Med Hyg 48:222–229

Laur F, Murgue B, Deparis X, Roche C, Cassar O, Chungue E (1998) Plasma levels of tumour necrosis factor alpha and transforming growth factor beta-1 in children with dengue 2 virus infection in French Polynesia. Trans R Soc Trop Med Hyg 92:654–656

Libraty DH, Pichyangkul S, Ajariyakhajorn C, Endy TP, Ennis FA (2001) Human dendritic cells are activated by dengue virus infection: enhancement by gamma interferon and implications for disease pathogenesis. J Virol 75:3501–3508

Libraty DH, Endy TP, Houng HS, Green S, Kalayanarooj S, Suntayakorn S, Chansiriwongs W, Vaughn DW, Nisalak A, Ennis FA et al (2002a) Differing influences of virus burden and immune activation on disease severity in secondary dengue-3 virus infections. J Infect Dis 185:1213–1221

Libraty DH, Young PR, Pickering D, Endy TP, Kalayanarooj S, Green S, Vaughn DW, Nisalak A, Ennis FA, Rothman AL (2002b) High circulating levels of the dengue virus nonstructural protein NS1 early in dengue illness correlate with the development of dengue hemorrhagic fever. J Infect Dis 186:1165–1168

Limonta D, Capo V, Torres G, Perez AB, Guzman MG (2007) Apoptosis in tissues from fatal dengue shock syndrome. J Clin Virol 40:50–54

Mitrakul C, Poshyachinda M, Futrakul P, Sangkawibha N, Ahandrik S (1977) Hemostatic and platelet kinetic studies in dengue hemorrhagic fever. Am J Trop Med Hyg 26:975–984

Mustafa AS, Elbishbishi EA, Agarwal R, Chaturvedi UC (2001) Elevated levels of interleukin-13 and IL-18 in patients with dengue hemorrhagic fever. FEMS Immunol Med Microbiol 30:229–233

Navarro-Sanchez E, Despres P, Cedillo-Barron L (2005) Innate immune responses to dengue virus. Arch Med Res 36:425–435

Nimmannitya S (1993) Clinical manifestations of Dengue/Dengue haemorrhagic fever. In: Thongcharoen P (ed) Monograph on Dengue/Dengue haemorrhagic fever. World Health Organization, India, pp 48–57

Pancharoen C, Rungsarannont A, Thisyakorn U (2002) Hepatic dysfunction in dengue patients with various severity. J Med Assoc Thai 85(Suppl 1):S298–301

Phanichyakarn P, Israngkura PB, Krisarin C, Pongpanich B, Dhanamitta S, Valyasevi A (1977) Studies on dengue hemorrhagic feverIV. Fluorescence staining of the immune complexes on platelets. J Med Assoc Thai 60:307–311

Pichyangkul S, Endy TP, Kalayanarooj S, Nisalak A, Yongvanitchit K, Green S, Rothman AL, Ennis FA, Libraty DH (2003) A blunted blood plasmacytoid dendritic cell response to an acute systemic viral infection is associated with increased disease severity. J Immunol 171:5571–5578

Raghupathy R, Chaturvedi UC, Al-Sayer H, Elbishbishi EA, Agarwal R, Nagar R, Kapoor S, Misra A, Mathur A, Nusrat H et al (1998) Elevated levels of IL-8 in dengue hemorrhagic fever. J Med Virol 56:280–285

Saito M, Oishi K, Inoue S, Dimaano EM, Alera MT, Robles AM, Estrella BD Jr, Kumatori A, Moji K, Alonzo MT et al (2004) Association of increased platelet-associated immunoglobulins with thrombocytopenia and the severity of disease in secondary dengue virus infections. Clin Exp Immunol 138:299–303

Shen BQ, Lee DY, Cortopassi KM, Damico LA, Zioncheck TF (2001) Vascular endothelial growth factor KDR receptor signaling potentiates tumor necrosis factor-induced tissue factor expression in endothelial cells. J Biol Chem 276:5281–5286

Sosothikul D, Seksarn P, Pongsewalak S, Thisyakorn U, Lusher J (2007) Activation of endothelial cells, coagulation and fibrinolysis in children with Dengue virus infection. Thromb Haemost 97:627–634

Srikiatkhachorn A, Ajariyakhajorn C, Endy TP, Kalayanarooj S, Libraty DH, Green S, Ennis FA, Rothman AL (2007) Virus-induced decline in soluble vascular endothelial growth receptor 2 is associated with plasma leakage in dengue hemorrhagic fever. J Virol 81:1592–1600

Suharti C, van Gorp EC, Dolmans WM, Setiati TE, Hack CE, Djokomoeljanto R, van der Meer JW (2003) Cytokine patterns during dengue shock syndrome. Eur Cytokine Netw 14:172–177

Sun DS, King CC, Huang HS, Shih YL, Lee CC, Tsai WJ, Yu CC, Chang HH (2007) Antiplatelet autoantibodies elicited by dengue virus non-structural protein 1 cause thrombocytopenia and mortality in mice. J Thromb Haemost 5:2291–2299

Tseng CS, Lo HW, Teng HC, Lo WC, Ker CG (2005) Elevated levels of plasma VEGF in patients with dengue hemorrhagic fever. FEMS Immunol Med Microbiol 43:99–102

Valero N, Larreal Y, Espina LM, Reyes I, Maldonado M, Mosquera J (2008) Elevated levels of interleukin-2 receptor and intercellular adhesion molecule 1 in sera from a venezuelan cohort of patients with dengue. Arch Virol 153:199–203

Vaughn DW, Green S, Kalayanarooj S, Innis BL, Nimmannitya S, Suntayakorn S, Endy TP, Raengsakulrach B, Rothman AL, Ennis FA et al (2000) Dengue viremia titer, antibody response pattern, and virus serotype correlate with disease severity. J Infect Dis 181:2–9

Wang S, He R, Patarapotikul J, Innis BL, Anderson R (1995) Antibody-enhanced binding of dengue-2 virus to human platelets. Virology 213:254–257

Wang L, Chen RF, Liu JW, Yu HR, Kuo HC, Yang KD (2007) Implications of dynamic changes among tumor necrosis factor-alpha (TNF-alpha), membrane TNF receptor, and soluble TNF receptor levels in regard to the severity of dengue infection. Am J Trop Med Hyg 77:297–302

Zarco RM, Espiritu-Campos L, Chan V (1957) Preliminary report on laboratory studies on Philippine hemorrhagic fever. J Philipp Med Assoc 33:676–683

Cellular Immunology of Sequential Dengue Virus Infection and its Role in Disease Pathogenesis

Alan L. Rothman

Contents

Abstract The increased risk for severe dengue disease during secondary dengue virus (DENV) infections, along with the clinical and pathological evidence pointing to cytokines as the proximal mediators of disease, has interested cellular immunologists in exploring the role of DENV-specific T lymphocytes in the pathogenesis of dengue-associated plasma leakage. Recent technological advances in the analysis

A.L. Rothman

Center for Infectious Disease and Vaccine Research, University of Massachusetts Medical School,
Worcester, MA, 01655, USA
e-mail: alan.rothman@umassmed.edu

A.L. Rothman (ed.), *Dengue Virus*, Current Topics in Microbiology and Immunology 338,   83
DOI 10.1007/978-3-642-02215-9_7, © Springer-Verlag Berlin Heidelberg 2010

of virus-specific T cells, applied to blood samples collected before, during and after acute DENV infections, have demonstrated that memory DENV-specific T cells from a prior infection respond with altered cytokine production to heterologous DENV serotypes and that the level of activation and expansion of these memory cells during acute DENV infection correlates with disease severity. Limited data suggest that specific T cell response profiles may predict a higher risk for severe disease during secondary infection. Application of these techniques and principles to animal models and to clinical trials may advance the development of novel therapeutics and protective vaccines.

1 Background

1.1 Hypothesis of T cell-mediated Immunopathogenesis

Early models of antibody-dependent enhancement recognized a missing factor – the link between increased DENV infection and disease pathogenesis. Halstead proposed "immune elimination" of DENV-infected cells as a key step in induction of plasma leakage (Halstead 1989), since viral infection was postulated to involve predominantly monocytic cells whereas plasma leakage reflected dysfunction of (uninfected) vascular endothelial cells. Additionally, clinicopathological observations such as the rapid onset and resolution of plasma leakage and the lack of ultrastructural damage to the vascular endothelium suggested that circulating mediators rather than direct viral effects were the proximate cause of endothelial dysfunction.

In parallel, work in animal models of other viral diseases, such as influenza, provided evidence for an important role of virus-specific T lymphocytes in both viral clearance and disease pathogenesis (Askonas et al. 1988). The ability of virus-specific T cells to lyse virus-infected cells and to secrete a wide range of cytokines fulfilled the characteristics of the "immune elimination" mechanism envisioned by Halstead. Based on these findings, Kurane and Ennis (Kurane and Ennis 1992) proposed that reactivation of memory DENV-specific T cells was an important element of the pathogenetic cascade leading to plasma leakage in DHF during secondary DENV infections.

1.2 Initial Evidence Supporting the Model of T cell Immunopathogenesis of DHF

A fundamental prediction of the model of T cell immunopathogenesis was that primary DENV infection induced memory T cells that were cross-reactive with heterologous DENV serotypes. Initial research efforts therefore focused on developing appropriate techniques and characterizing the specificity of T cell responses to DENV. Because of the need for large volumes of blood to perform most assays of

T cell specificity and function and the difficulty in defining the history of prior DENV infections in adults from most dengue-endemic countries, much of this early characterization of DENV-specific T lymphocytes was performed using adult volunteers who were injected with monovalent live candidate DENV vaccine strains (Kurane et al. 1989b).

Peripheral blood mononuclear cells from these donors, tested months to years after immunization, showed proliferation and interferon-γ (predominantly CD4 T cell) responses (Kurane et al. 1989a) as well as cytolytic (predominantly CD8 T cell) responses (Mathew et al. 1996) to the DENV serotype with which the donors were immunized. Responses to heterologous DENV serotypes were variable among donors but cross-reactivity to one or more heterologous serotypes was detected in all cases. The isolation of DENV-reactive CD4 and CD8 T cell clones provided a detailed explanation for serotype-crossreactivity (Kurane et al. 1989b). The DENV-reactive memory T cell repertoire in several individuals was demonstrated to contain both clones specific for the immunizing serotype as well as clones cross-reactive with one, two, or all three heterologous DENV serotypes; T cell clones cross-reactive with other flaviviruses, such as West Nile virus, were also identified (Kurane et al. 1995). Single amino acid changes within the epitope recognized were typically responsible for the fine specificity of T cell recognition of different DENV serotypes and several examples of recognition of the same epitope by both serotype-specific and serotype-crossreactive cells have been described (Livingston et al. 1995).

A second key prediction of the model of T cell-mediated immunopathogenesis was that patients with DHF would have greater T cell activation and T cell cytokine production than patients with less severe disease. This prediction was tested using various cohorts of patients with natural DENV infections. Several studies demonstrated that serum samples collected from patients hospitalized with DHF had elevated levels of various cytokines, including IFN-γ, IL-2 and TNFα, as well as soluble markers of immune activation such as sCD4, sCD8, sIL-2R and sTNFR, compared to normal controls or nonhospitalized patients with dengue (Green et al. 1999b; Hober et al. 1996; Hober et al. 1993; Kurane et al. 1991).

1.3 Gaps in Knowledge

Assays of DENV-specific T cell proliferation and cytolytic activity, performed on whole PBMC samples, provided an imprecise measure of the magnitude of the response in each individual and reflected the sum of many different responses. Data from T cell clones provided additional detail but were difficult to extrapolate to the overall response to secondary DENV infection. Furthermore, neither of these methods yielded data on the state of virus-specific T cells in vivo during acute infection. Cytokine levels in patient sera reflected overall production from many different cell types and changed rapidly over the course of illness. Differences in timing and method of blood collection from different patient cohorts confounded the interpretation of the differences in cytokine levels detected. Several key pieces of the puzzle of DHF were clearly missing from these analyzes. First, it was necessary to determine which epitope-specific T cell responses made the greatest contribution to the overall immune response (immunodominance). Second, it was necessary to determine how the T cells responded in vivo during acute DENV infection. Lastly, well-characterized patient cohorts would be needed to facilitate interpretation of the immunological assays.

The model of T cell immunopathogenesis presented a theoretical basis for use of immunomodulatory drugs in treatment of established DHF, or in early treatment of patients with dengue to minimize the development of plasma leakage. Thus, confirmation of this model had important therapeutic implications. Nevertheless, an important long-range goal of immunologic studies was to distinguish protective from pathological immune responses in natural DENV infections in order to guide the development of effective vaccines. To accomplish that goal, it was also necessary to establish prospective cohorts and to collect PBMC prior to secondary DENV infections in order to identify the immunological predictors (or correlates) of benign versus pathological clinical outcomes.

2 Methodologic Advances

2.1 *Immunologic Techniques*

Beginning in the late 1990s, new methods for analysis of virus-specific T cells set off a revolution in immunology. ELISPOT and flow cytometry techniques to detect individual cytokine-producing cells established the ability to accurately quantify the frequencies of virus-specific T cells (Murali-Krishna et al. 1998). A novel reagent, tetrameric complexes of peptide-loaded MHC molecules, provided the capability of detecting antigen-specific T cells without requiring a specific functional response (Altman et al. 1996). Of critical importance, these methods proved applicable even during acute phase of viral infection. In murine infection with LCMV and human infection with HIV and EBV, these methods showed that the frequencies of virus-specific T cells were orders of magnitude higher than had been estimated based on bulk culture methods and limiting dilution cloning (Catalina et al. 2002; Murali-Krishna et al. 1998; Tan et al. 1999). These advances were possible because they built upon many prior studies defining immunodominant viral epitopes, particularly in established mouse models of acute viral infection. It remained unclear, however, how generalizable these findings would be to other acute viral infections in humans.

2.2 *Databases*

The maturation of automated genomic amplification and sequencing methods during the 1990s also advanced progress in studying DENV immunology through an expansion of databases of viral sequence information. In combination with greater commercial availability of synthetic peptides, higher-throughput immuno-logic assays and improved methods for selection of candidate epitopes, this sequence information facilitated the screening of the DENV polyprotein for immuno-dominant epitopes. Given the diversity of HLA alleles in the human population and the accessibility of the entire DENV polyprotein to HLA class I and II antigen presentation pathways, lack of knowledge of the important regions of the genome had previously been an obstacle to characterization of T lymphocyte responses to DENV.

2.3 Study Populations

Most previous clinical studies of dengue disease and immune responses relied on "convenience" cohorts, typically DHF patients identified after admission to the hospital and DF patients seen in outpatient clinics during disease outbreaks. These patient populations were very heterogeneous with regard to duration of illness prior to study and were usually enrolled later in illness, when the success rate for identifying the infecting DENV serotype was low. These factors clouded the interpretation of the immunologic data; comparison groups drawn from different settings (e.g., comparing outpatients to inpatients) introduced additional confoun-ders. Furthermore, the technical expertise of staff and blood volumes required for PBMC isolation had not been available to most investigators.

In order to address the fundamental cellular immunology of acute DENV infection, we and others recognized the need to establish prospective study cohorts specifically for immunologic investigations. The Dengue Hemorrhagic Fever Proj-ect, a collaboration among the University of Massachusetts, the Armed Forces Research Institute of Medical Sciences and the Queen Sirikit National Institute of Child Health in Bangkok, Thailand, established a cohort of children with acute dengue illness (Kalayanarooj et al. 1997; Vaughn et al. 1997). Key elements of the study design included enrollment at a very early stage of disease (within 72 h of onset), daily blood collection with isolation of PBMC and detailed assessment of disease severity. These design features provided a very high frequency of detectable viremia, well-matched DF and DHF groups and serial blood samples illustrating the changes in immune responses over time. Similar prospective studies of acute dengue illness have been established by other groups working in Vietnam, Nicaragua and elsewhere (Hammond et al. 2005; Wills et al. 2002)

Our group, in collaboration with AFRIMS and the Ministry of Public Health, Thailand, also established a separate population-based cohort study in Kamphaeng Phet province, to study the risk factors for mild vs. severe dengue illness (Endy et al. 2002a; Endy et al. 2002b). The Kamphaeng Phet study prospectively enrolled over 2,000 primary school children and defined the incidence of clinical and subclinical DENV infections over a 5-year period. Serum and PBMC collected annually from the entire cohort has provided a unique resource to explore the immunological correlates of protection versus pathology. Intense active surveillance for DENV infections has also defined the circulation of each of the four DENV serotypes in this population, as described further elsewhere in this volume.

2.4   Funding

The disproportionate burden of dengue on resource-limited countries has contributed to its status among the "neglected tropical diseases," as well as the historically low number of investigators from developed countries involved in research in this field. The Military Infectious Disease Research Program has, for many decades, been a lonely bright spot in the field, providing core funding for overseas laboratories such as the AFRIMS laboratory in Thailand. That picture has changed significantly in the last 10–20 years. NIH support for research on dengue has increased substantially, especially in conjunction with the biodefense initiatives that have been mounted since 2001. The Wellcome Trust has also invested significant resources to establish field research in the developing world and train new investigators in research on tropical diseases, including dengue. The Pediatric Dengue Vaccine Initiative, supported by a major commitment of funds from the Bill and Melinda Gates Foundation, has established an ambitious agenda of basic and applied research towards the goal of introducing effective dengue vaccine(s) into endemic countries. More recently, the European Union has funded two large, multinational projects on dengue in coordination with the WHO. These efforts have already had and will likely continue to have, a large impact on scientific progress.

3   Recent Observations

3.1   Application of New Methods to the Study of DENV-specific T cell Responses

We and other laboratories have successfully applied these new immunological techniques – ELISPOT (Simmons et al. 2005; Zivna et al. 2002), cytokine flow cytometry (Bashyam et al. 2006; Dong et al. 2007; Guy et al. 2008; Imrie et al. 2007; Mangada and Rothman 2005) and HLA-peptide tetramer staining (Dong et al. 2007; Mongkolsapaya et al. 2003; Mongkolsapaya et al. 2006) – to study DENV-specific T cell responses. Most such assays have employed synthetic peptides based on prototype DENV sequences and have characterized memory CD8 T cell responses. We have also used inactivated lysates of DENV-infected cells in cytokine flow cytometry assays; inactivated antigens poorly induce responses in CD8 T cells, however. The choice of antigen for these studies presents several challenges. First, a large number of synthetic peptides is required for analysis of the full DENV polyprotein (~3,000 amino acids in length), particularly given the existence of four distinct serotypes. Given the cost, both in synthesis of reagents and in PBMC required, most groups have been obligated to study a subset of all potential epitopes. Secondly, when using synthetic peptides, viruses, or inactivated cell lysates, not all potential viral variants can be studied. Most groups have focused on prototype DENV strains that do not necessarily reflect current circulating viruses. Although amino acid sequence variation within a given DENV serotype is small, these sequence differences can affect the T cell response. To address these challenges, the NIH has recently established a repository of synthetic peptides, including peptides from multiple DENV strains.

The experience with the newer immunological techniques to date has confirmed several critical observations. First, these assays have demonstrated a high frequency of DENV-specific memory T cells, in some cases encompassing over 1% of circulating T cells (Bashyam et al. 2006; Mongkolsapaya et al. 2003). Secondly, DENV-specific T cells have been demonstrated on a single-cell level to display various effector functions, including degranulation, production of multiple cytokines including IFN-γ, TNFα, MIP-1β and IL-2 and proliferation in response to stimulation with the cognate DENV antigen (Bashyam et al. 2006; Dong et al. 2007; Imrie et al. 2007). Thirdly, it has been possible to directly demonstrate a high frequency of T cells that recognize heterologous DENV serotypes (Bashyam et al. 2006; Mongkolsapaya et al. 2003).

3.2   Identification of DENV T cell Epitopes

The recent studies have identified a number of novel epitopes recognized by DENV-specific CD4 and CD8 T cells (Bashyam et al. 2006; Imrie et al. 2007; Mangada and Rothman 2005; Mongkolsapaya et al. 2003; Mongkolsapaya et al. 2006; Simmons et al. 2005; Zivna et al. 2002). Several of these epitopes appear to be immunodominant, i.e., recognized by T cells present at high frequency in a majority of individuals who share the restricting HLA allele. Such epitopes are of particular interest for studies of disease associations, as cohorts of patients can potentially be studied using a limited number of reagents. However, this presumes that responses to these peptides reflect the overall T cell response in vivo. Particular HLA alleles have been associated (positively or negatively) with more severe dengue disease (Loke et al. 2001; Stephens et al. 2002), as discussed in greater detail elsewhere in this volume. Unfortunately, most of the T cell epitopes defined to date are not recognized in the context of the HLA alleles with known disease association.

Immunodominant T cell epitopes have been identified on multiple DENV proteins. Interestingly, many of these epitopes have been identified on the nonstructural protein NS3, consistent with earlier findings using older immunologic methods. This conclusion may be biased, however; as noted above, many studies did not test a comprehensive set of DENV peptides and the earlier findings led many groups to focus their studies on peptides in the NS3 protein. Most of the immunodominant epitopes have sequence differences between the four DENV serotypes, again confirming observations made in earlier studies with T cell clones.

3.3 Altered Peptide Ligands

We originally proposed a model of altered functional responses of DENV-specific
T cells to heterologous viral serotypes based on results using T cell clones (Zivny
et al. 1999). The recent studies of Imrie et al (Imrie et al. 2007) and Dong et al
(Dong et al. 2007) have shown similar results with T cell clones specific for
epitopes on NS5 and NS3, respectively. In addition, their studies demonstrated
that the distinct T cell functional responses to heterologous serotypes corresponded
with differences in avidity to the variant peptide sequences.

We analyzed T cell cytokine responses in PBMC obtained following primary
DENV infection by intracellular cytokine staining (Mangada and Rothman 2005).
CD4 T cell responses to inactivated DENV-infected cell lysates (containing all DENV
proteins) demonstrated serotype-crossreactive IFNγ and TNFα responses; however,
the ratio of TNFα- to IFNγ-producing cells was higher in response to heterologous
DENV antigens than homologous DENV antigens. CD4 T cell responses to single
epitopes showed a more variable pattern. CD8 T cell responses to DENV peptides also
demonstrated cross-reactive IFNγ, TNFα and MIP-1β responses to heterologous
serotypes (Bashyam et al. 2006). We detected subpopulations of T cells expressing
all possible combinations of these cytokines and both the total frequency of responding cells and the distribution of specific patterns of cytokine production varied across
epitopes and between donors (for the same epitope). Interestingly, we found that some
heterologous DENV peptides were stronger inducers of cytokine production than the
corresponding homologous peptide.

These studies convincingly demonstrate that T cell responses to heterologous
DENV serotypes reflect not only quantitative differences but qualitative differences
as well. Furthermore, in addition to altered cytokine production due to lower T cell
avidity for heterologous peptides, these results indicate that some peptide variants
may actually act as "superagonists," inducing more robust cytokine production. The
results suggest that the overall T cell functional response to secondary DENV
infection will likely depend upon both HLA type and sequence of infection with
different serotypes but may still show significant variation between donors.

3.4 Acute Infection- Expansion, Timing, Relationship to Disease

While studies of DENV-specific memory T cell populations have led to hypotheses
about their function during secondary infection, it is essential to confirm these
hypotheses through observation of T cell activation in vivo during acute DENV
infection. These studies have been greatly facilitated by the tools described above.
Initial studies from our group showed high levels of CD8 T cell activation as well as
evidence of oligoclonal T cell expansions based on T cell receptor Vβ gene usage in
acute secondary DENV infection (Gagnon et al. 2001). T cell activation, as
measured by the percentage of CD69+ T cells, was higher in patients with DHF,

consistent with a causal relationship to severe disease (Green et al. 1999a); however, these studies did not specifically identify DENV-specific T cells.

Using both IFNγ ELISPOT assays and HLA-peptide tetramer staining, we and others have more recently demonstrated that DENV epitope-specific T cells make up a very high frequency (up to ~8%) of circulating CD8 T cells during and in the 1–2 weeks after acute secondary DENV infection (Dong et al. 2007; Mongkolsapaya et al. 2003; Mongkolsapaya et al. 2006; Simmons et al. 2005; Zivna et al. 2002). Several studies confirmed that more severe dengue disease was associated with higher frequencies of DENV-specific T cells (Mongkolsapaya et al. 2003; Zivna et al. 2002). Mongkolsapaya et al reported that the frequency of tetramer-positive cells reached a maximum 2 weeks after acute infection; however, our results studying a different epitope found that peak T cell frequencies were achieved during the febrile (and viremic) phase (unpublished data), more consistent with their proposed role in pathogenesis.

T cell frequencies detected by HLA-peptide tetramers have been considerably higher than those detected in ELISPOT assays or intracellular cytokine staining. Mongkolsapaya et al have suggested that this "suboptimal" functional response of T cells in secondary DENV infection may contribute to immunopathology (Mongkolsapaya et al. 2006). However, this may instead reflect a deficit in in vitro functional responses of cells recently activated in vivo, as has been found in other acute viral infections (Lechner et al. 2000). Using double tetramer staining, Mongkolsapaya et al also showed greater staining with tetramers containing peptides of heterologous DENV serotypes (Mongkolsapaya et al. 2003). This finding is consistent with recall of T cell responses induced by an earlier primary DENV infection. However, the serotype causing the primary infection was not known for these patients and the alternative possibility that the pattern of tetramer staining reflected the inherent immunogenicity of each serotype could not be excluded. Our data show substantial differences between subjects in the pattern of staining with homologous and heterologous tetramers (unpublished data) and suggest that the baseline frequency of epitope-specific T cells may affect the degree to which the immunodominance effect skews the response to secondary infection. Resolution of these possibilities will need to await further clinical studies.

3.5 Association of Preexisting Responses to Disease During Secondary Infection

The characteristics of the T cell response during secondary DENV infection are likely influenced by both the preexisting DENV-specific T cell repertoire and early events in secondary infection, such as initial viral burden and innate immune responses. Prospective studies in populations at risk for DENV infection are the only means available to draw these associations. We have begun to analyze such associations in the Kamphaeng Phet cohort described above (Mangada et al. 2002).

T cell proliferation assays using DENV-infected cell lysates did not show clear associations with the need for hospitalization (or lack thereof). Interestingly, TNFα production in these cultures was only detected among children who were hospitalized during their subsequent secondary infection. IFNγ secretion did not show a significant association with disease but there was a trend toward more serotype-crossreactive IFNγ production among children with milder disease. These preliminary findings suggest that there may be multiple profiles of DENV-specific T cell responses associated with either increased or decreased risk for severe dengue disease. However, further studies are needed to confirm and extend these findings.

We have also begun to study the Kamphaeng Phet cohort using intracellular cytokine staining. No clear disease associations were seen with the frequency of IFNγ-producing T cells. Interestingly, overall we observed increases in T cell frequencies to the homologous DENV serotype following secondary infection, whereas T cell frequencies to heterologous serotypes were more variable, with some individuals showing increases and others decreases (unpublished data). These findings point to marked changes in the DENV-specific T cell repertoire as a result of secondary infection.

3.6 T cell Responses to Tetravalent Vaccination

The principles derived from studies of natural and experimental infection with individual DENV viruses are only beginning to be applied to characterize the T cell response induced by tetravalent live attenuated DENV vaccines (Guy et al. 2008; Gwinn et al. 2003; Rabablert et al. 2000; Rothman et al. 2001). Studies to date have shown the induction of T cell proliferation, IFNγ production and cytolytic responses to multiple DENV serotypes similar to the responses induced by natural infection, including authentic serotype-crossreactive T cells. Responses have not been equivalent to all four DENV serotypes; however, this may reflect interference between the four vaccine viruses, as has been reflected in viremia levels and neutralizing antibody titers and most of these vaccines have been modified for further clinical development. Without information on the clinical efficacy of the vaccines, the implications of the findings on cellular immunity remain speculative.

3.7 Animal Models of Sequential DENV Infection

Our understanding of the role of DENV-specific T cell responses in secondary infections would be greatly advanced by animal models that allow equivalent experimental groups to be defined in a controlled manner. Although good models of dengue disease remain elusive, some progress has been made in characterizing the cellular immunology of sequential DENV infections in nonhuman primates (Chen et al. 2007; Koraka et al. 2007) and mice (Beaumier et al. 2008). While immunocompetent mice do not develop measurable DENV viremia, we found many similarities between the T cell response to heterologous secondary infection in mice and those observed in humans (Beaumier et al. 2008). Specifically, the T cell response was enhanced to a degree that was dependent on the sequence of DENV serotypes and some epitope variants were immunodominant regardless of the infecting serotype. More recently, we have detected human T cell responses to DENV after infection of "humanized" mice- immunodeficient mice reconstituted with human hematopoietic stem cells (unpublished data). Further evaluation of these models is needed; however, these results suggest that some hypotheses regarding the effects of DENV-specific T cells can be tested experimentally in animal models.

4 Perspective

4.1 Model

The accumulating data continue to support an important role for DENV-specific memory T cells in the effector phase of disease pathogenesis in secondary DENV infections. Nevertheless, the picture continues to grow more complex (Fig. 1). Mosquito vector-human host interactions, viral determinants of cellular tropism and replication, host genetics, innate immune responses and preexisting serotype-crossreactive antibodies all affect the earliest ("afferent") phases of DENV infection. The adaptive cellular immune system is subsequently activated and may, under certain conditions, initiate "efferent" response cascades affecting vascular permeability and leading to clinically overt plasma leakage (DHF). The preexisting T cell repertoire capable of responding to the (secondary) infecting DENV serotype, which is modulated by host genetics (particularly HLA polymorphisms) as well as the serotype of the previous infection, determines the likelihood of a benign or pathological outcome. We are only beginning to define those characteristics of the T cell repertoire that have the greatest impact on the likelihood of disease.

4.2 Implications

Well-designed clinical studies have demonstrated that high dose glucocorticoids offer no benefit in patients with established shock (Panpanich et al. 2006; Tassniyom et al. 1993). Therefore, while immunomodulatory drugs represent an attractive approach to therapy based on the immunopathological mechanisms underlying DHF, this cannot be recommended outside the research setting and should only be done in carefully designed clinical trials.

It can be anticipated that the live attenuated vaccines currently in advanced stages of development will induce DENV-specific memory CD4 and CD8 T cells. The induction of robust T cell responses may be important for long-term protective

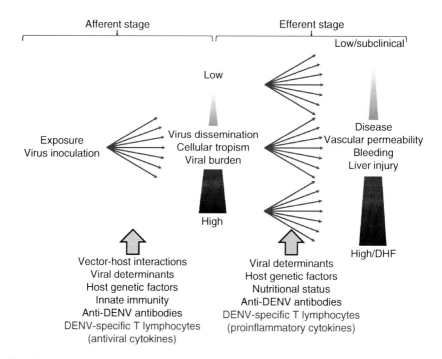

Fig. 1 Proposed model of the steps leading from initial dengue virus (DENV) infection to development of dengue disease. The process begins from the point of intradermal/subcutaneous virus inoculation by an infected Aedes mosquito. Initial viral replication at the site of inoculation and/or regional lymphatics leads to systemic dissemination (afferent stage). A spectrum of viral burden is observed in human populations; variation in cellular tropism is also possible, although there is little published data. Potential determinants of viral distribution and burden include vector-host interactions, viral determinants ("virulence"), host genetic factors, innate immunity and preexisting anti-DENV antibodies (e.g., antibody-dependent enhancement of infection). Clinical manifestations of dengue disease, including vascular permeability, bleeding and liver injury, result from systemic DENV infection (efferent stage). While viral burden has a strong influence on disease severity, additional factors affect this relationship, including viral determinants, host genetic factors, nutritional status and antiDENV antibodies. DENV-specific T lymphocytes, both naïve T cells and memory T cells (from prior flavivirus infections), are activated during acute infection and potentially influence both afferent and efferent stages. Early production of antiviral cytokines, such as interferon-γ, may inhibit viral replication and limit peak viral burden. Pro-inflammatory cytokines, such as tumor necrosis factor α (which can act synergistically with late interferon-γ production) may affect vascular permeability and contribute to enhanced disease

immunity; however, there is also the potential for vaccines to induce nonprotective, pathological immune responses. Furthermore, simultaneous immunization with multiple viruses may not have the same effect as sequential infection and may modify the pattern of immunodominance (Zhou and Deem 2006). As we are not currently able to reliably distinguish desirable and undesirable responses, these considerations should not be an obstacle to clinical trials. Nevertheless, it would be important for the first vaccine efficacy studies to provide further insights to guide later efforts.

4.3 Research Needs

While substantial progress has been made in recent years, significant gaps remain in our understanding of DENV-specific T lymphocyte responses and their associations with disease. I offer the following broad suggestions for further research:

- Primary infection – What is the course of activation of naïve DENV-specific T cells during acute primary DENV infection? How does it differ from secondary infection? Are the same immunologic associations seen in DHF occurring during primary infections?
- Immunodominance – What are the principles that determine immunodominance of particular DENV epitopes (and serotypes)? How is this affected by the sequence of infections? Is immunodominance a critical factor in the associations between T lymphocyte responses and disease?
- Animal models – How can animal models be improved to reproduce the immunological response in humans to secondary DENV infection? Will improved animal models recapitulate DHF? Can this information be used to test vaccine strategies?
- Immunological correlates – What patterns of T lymphocyte responses are most strongly correlated with protection from severe dengue disease? What patterns are most strongly correlated with enhanced risk for severe dengue disease? Do the same correlates apply for vaccine-induced T cell responses as for natural infection? Do the same correlates apply for different vaccines? Can T cell responses be used to predict the clinical results with new vaccines?
- Vaccination – What are the characteristics (serotype-crossreactivity, effector responses) of the T lymphocyte response to tetravalent vaccines? Does immunization with chimeric flaviviruses alter the specificity of the T lymphocyte response? How does the immunization method (dose, location, timing and sequence) affect the resulting T lymphocyte response? Can vaccine administration be manipulated to "sculpt" the immune response to achieve a desired result?

References

Altman JD, Moss PA, Goulder PJ, Barouch DH, McHeyzer-Williams MG, Bell JI, McMichael AJ, Davis MM (1996) Phenotypic analysis of antigen-specific T lymphocytes. Science 274:94–96

Askonas BA, Taylor PM, Esquivel F (1988) Cytotoxic T cells in influenza infection. Ann N Y Acad Sci 532:230–237

Bashyam HS, Green S, Rothman AL (2006) Dengue virus-reactive CD8+ T cells display quantitative and qualitative differences in their response to variant epitopes of heterologous viral serotypes. J Immunol 176:2817–2824

Beaumier CM, Mathew A, Bashyam HS, Rothman AL (2008) Cross-reactive memory CD8(+) T cells alter the immune response to heterologous secondary dengue virus infections in mice in a sequence-specific manner. J Infect Dis 197:608–617

Catalina MD, Sullivan JL, Brody RM, Luzuriaga K (2002) Phenotypic and functional heterogeneity of EBV epitope-specific CD8+ T cells. J Immunol 168:4184–4191

Chen L, Ewing D, Subramanian H, Block K, Rayner J, Alterson KD, Sedegah M, Hayes C, Porter K, Raviprakash K (2007) A heterologous DNA prime-Venezuelan equine encephalitis virus replicon particle boost dengue vaccine regimen affords complete protection from virus challenge in cynomolgus macaques. J Virol 81:11634–11639

Dong T, Moran E, Vinh Chau N, Simmons C, Luhn K, Peng Y, Wills B, Phuong Dung N, Thi Thu Thao L, Hien TT, McMichael A, Farrar J, Rowland-Jones S (2007) High pro-inflammatory cytokine secretion and loss of high avidity cross-reactive cytotoxic T-cells during the course of secondary dengue virus infection. PLoS ONE 2:e1192

Endy TP, Chunsuttiwat S, Nisalak A, Libraty DH, Green S, Rothman AL, Vaughn DW, Ennis FA (2002a) Epidemiology of inapparent and symptomatic acute dengue virus infection: a prospective study of primary school children in Kamphaeng Phet, Thailand. Am J Epidemiol 156:40–51

Endy TP, Nisalak A, Chunsuttiwat S, Libraty DH, Green S, Rothman AL, Vaughn DW, Ennis FA (2002b) Spatial and temporal circulation of dengue virus serotypes: a prospective study of primary school children in Kamphaeng Phet, Thailand. Am J Epidemiol 156:52–59

Gagnon SJ, Leporati A, Green S, Kalayanarooj S, Vaughn DW, Stephens HA, Suntayakorn S, Kurane I, Ennis FA, Rothman AL (2001) T cell receptor Vbeta gene usage in Thai children with dengue virus infection. Am J Trop Med Hyg 64:41–48

Green S, Pichyangkul S, Vaughn DW, Kalayanarooj S, Nimmannitya S, Nisalak A, Kurane I, Rothman AL, Ennis FA (1999a) Early CD69 expression on peripheral blood lymphocytes from children with dengue hemorrhagic fever. J Infect Dis 180:1429–1435

Green S, Vaughn DW, Kalayanarooj S, Nimmannitya S, Suntayakorn S, Nisalak A, Lew R, Innis BL, Kurane I, Rothman AL, Ennis FA (1999b) Early immune activation in acute dengue is related to development of plasma leakage and disease severity. J Infect Dis 179:755–762

Guy B, Nougarede N, Begue S, Sanchez V, Souag N, Carre M, Chambonneau L, Morrisson DN, Shaw D, Qiao M, Dumas R, Lang J, Forrat R (2008) Cell-mediated immunity induced by chimeric tetravalent dengue vaccine in naive or flavivirus-primed subjects. Vaccine 26:5712–5721

Gwinn W, Sun W, Innis BL, Caudill J, King AD (2003) Serotype-specific T(H)1 responses in recipients of two doses of candidate live-attenuated dengue virus vaccines. Am J Trop Med Hyg 69:39–47

Halstead SB (1989) Antibody, macrophages, dengue virus infection, shock, and hemorrhage: a pathogenetic cascade. Rev Infect Dis 11:S830–S839

Hammond SN, Balmaseda A, Perez L, Tellez Y, Saborio SI, Mercado JC, Videa E, Rodriguez Y, Perez MA, Cuadra R, Solano S, Rocha J, Idiaquez W, Gonzalez A, Harris E (2005) Differences in dengue severity in infants, children, and adults in a 3-year hospital-based study in Nicaragua. Am J Trop Med Hyg 73:1063–1070

Hober D, Delannoy AS, Benyoucef S, De Groote D, Wattre P (1996) High levels of sTNFR p75 and TNF alpha in dengue-infected patients. Microbiol Immunol 40:569–573

Hober D, Poli L, Roblin B, Gestas P, Chungue E, Granic G, Imbert P, Pecarere JL, Vergez-Pascal R, Wattre P, Maniez-Montreuil M (1993) Serum levels of tumor necrosis factor-α (TNF-α), interleukin-6 (IL-6), and interleukin-1β (IL-1β) in dengue-infected patients. Am J Trop Med Hyg 48:324–331

Imrie A, Meeks J, Gurary A, Sukhbataar M, Kitsutani P, Effler P, Zhao Z (2007) Differential functional avidity of dengue virus-specific T-cell clones for variant peptides representing heterologous and previously encountered serotypes. J Virol 81:10081–10091

Kalayanarooj S, Vaughn DW, Nimmannitya S, Green S, Suntayakorn S, Kunentrasai N, Viramitrachai W, Ratanachu-eke S, Kiatpolpoj S, Innis BL, Rothman AL, Nisalak A, Ennis FA (1997) Early clinical and laboratory indicators of acute dengue illness. J Infect Dis 176:313–321

Koraka P, Benton S, van Amerongen G, Stittelaar KJ, Osterhaus AD (2007) Efficacy of a live attenuated tetravalent candidate dengue vaccine in naive and previously infected cynomolgus macaques. Vaccine 25:5409–5416

Kurane I, Ennis FA (1992) Immunity and immunopathology in dengue virus infections. Sem Immunol 4:121–127

Kurane I, Innis BL, Nimmannitya S, Nisalak A, Meager A, Janus J, Ennis FA (1991) Activation of T lymphocytes in dengue virus infections. High levels of soluble interleukin 2 receptor, soluble CD4, soluble CD8, interleukin 2, and interferon-gamma in sera of children with dengue. J Clin Invest 88:1473–1480

Kurane I, Innis BL, Nisalak A, Hoke C, Nimmannitya S, Meager A, Ennis FA (1989a) Human T cell responses to dengue virus antigens. Proliferative responses and interferon gamma production. J Clin Invest 83:506–513

Kurane I, Meager A, Ennis FA (1989b) Dengue virus-specific human T cell clones. Serotype crossreactive proliferation, interferon gamma production, and cytotoxic activity. J Exp Med 170:763–775

Kurane I, Okamoto Y, Dai LC, Zeng LL, Brinton MA, Ennis FA (1995) Flavivirus-cross-reactive, HLA-DR15-restricted epitope on NS3 recognized by human CD4+ CD8- cytotoxic T lymphocyte clones. J Gen Virol 76:2243–2249

Lechner F, Wong DK, Dunbar PR, Chapman R, Chung RT, Dohrenwend P, Robbins G, Phillips R, Klenerman P, Walker BD (2000) Analysis of successful immune responses in persons infected with hepatitis C virus. J Exp Med 191:1499–1512

Livingston PG, Kurane I, Dai LC, Okamoto Y, Lai CJ, Men R, Karaki S, Takiguchi M, Ennis FA (1995) Dengue virus-specific, HLA-B35-restricted, human CD8+ cytotoxic T lymphocyte (CTL) clones. Recognition of NS3 amino acids 500 to 508 by CTL clones of two different serotype specificities. J Immunol 154:1287–1295

Loke H, Bethell DB, Phuong CX, Dung M, Schneider J, White NJ, Day NP, Farrar J, Hill AV (2001) Strong HLA class I–restricted T cell responses in dengue hemorrhagic fever: a double-edged sword? J Infect Dis 184:1369–1373

Mangada MM, Endy TP, Nisalak A, Chunsuttiwat S, Vaughn DW, Libraty DH, Green S, Ennis FA, Rothman AL (2002) Dengue-specific T cell responses in peripheral blood mononuclear cells obtained prior to secondary dengue virus infections in Thai schoolchildren. J Infect Dis 185:1697–1703

Mangada MM, Rothman AL (2005) Altered cytokine responses of dengue-specific CD4+ T cells to heterologous serotypes. J Immunol 175:2676–2683

Mathew A, Kurane I, Rothman AL, Zeng LL, Brinton MA, Ennis FA (1996) Dominant recognition by human CD8+ cytotoxic T lymphocytes of dengue virus nonstructural proteins NS3 and NS1.2a. J Clin Invest 98:1684–1694

Mongkolsapaya J, Dejnirattisai W, Xu X, Vasanawathana S, Tangthawornchaikul N, Chairunsri A, Sawasdivorn S, Duangchinda T, Dong T, Rowland-Jones S, Yenchitsomanus P, McMichael A, Malasit P, Screaton G (2003) Original antigenic sin and apoptosis in the pathogenesis of dengue hemorrhagic fever. Nature Med 9(7):921–927

Mongkolsapaya J, Duangchinda T, Dejnirattisai W, Vasanawathana S, Avirutnan P, Jairungsri A, Khemnu N, Tangthawornchaikul N, Chotiyarnwong P, Sae-Jang K, Koch M, Jones Y, McMichael A, Xu X, Malasit P, Screaton G (2006) T cell responses in dengue hemorrhagic fever: are cross-reactive T cells suboptimal? J Immunol 176:3821–3829

Murali-Krishna K, Altman JD, Suresh M, Sourdive DJ, Zajac AJ, Miller JD, Slansky J, Ahmed R (1998) Counting antigen-specific CD8 T cells: a reevaluation of bystander activation during viral infection. Immunity 8:177–187

Panpanich R, Sornchai P, Kanjanaratanakorn K (2006) Corticosteroids for treating dengue shock syndrome. Cochrane Database Syst Rev 3:CD003488

Rabablert J, Dharakul T, Yoksan S, Bhamarapravati N (2000) Dengue virus specific T cell responses to live attenuated monovalent dengue-2 and tetravalent dengue vaccines. Asian Pac J Allergy Immunol 18:227–35

Rothman AL, Kanesa-thasan N, West K, Janus J, Saluzzo J, Ennis FA (2001) Induction of T lymphocyte responses to dengue virus by a candidate tetravalent live attenuated dengue virus vaccine. Vaccine 19:4694–9

Simmons CP, Dong T, Chau NV, Dung NT, Chau TN, Thao-le TT, Hien TT, Rowland-Jones S, Farrar J (2005) Early T-cell responses to dengue virus epitopes in Vietnamese adults with secondary dengue virus infections. J Virol 79:5665–5675

Stephens HA, Klaythong R, Sirikong M, Vaughn DW, Green S, Kalayanarooj S, Endy TP, Libraty
    DH, Nisalak A, Innis BL, Rothman AL, Ennis FA, Chandanayingyong D (2002) HLA-A and
    -B allele associations with secondary dengue virus infections correlate with disease severity
    and the infecting viral serotype in ethnic Thais. Tissue Antigens 60:309–18
Tan LC, Gudgeon N, Annels NE, Hansasuta P, O'Callaghan CA, Rowland-Jones S, McMichael
    AJ, Rickinson AB, Callan MF (1999) A re-evaluation of the frequency of CD8+ T cells specific
    for EBV in healthy virus carriers. J Immunol 162:1827–35
Tassniyom S, Vasanawathana S, Chirawatkul A, Rojanasuphot S (1993) Failure of high-dose
    methylprednisolone in established dengue shock syndrome: a placebo-controlled, double-blind
    study. Pediatrics 92:111–115
Vaughn DW, Green S, Kalayanarooj S, Innis BL, Nimmannitya S, Suntayakorn S, Rothman AL,
    Ennis FA, Nisalak A (1997) Dengue in the early febrile phase: viremia and antibody responses.
    J Infect Dis 176:322–330
Wills BA, Oragui EE, Stephens AC, Daramola OA, Dung NM, Loan HT, Chau NV, Chambers M,
    Stepniewska K, Farrar JJ, Levin M (2002) Coagulation abnormalities in dengue hemorrhagic
    fever: serial investigations in 167 Vietnamese children with Dengue shock syndrome. Clin
    Infect Dis 35:277–85
Zhou H, Deem MW (2006) Sculpting the immunological response to dengue fever by polytopic
    vaccination. Vaccine 24:2451–9
Zivna I, Green S, Vaughn DW, Kalayanarooj SH, Stephens AF, Chandanayingyong D, Nisalak A,
    Ennis FA, Rothman AL (2002) T cell responses to an HLA B*07-restricted epitope on the
    dengue NS3 protein correlate with disease severity. J Immunol 168:5959–5965
Zivny J, DeFronzo M, Jarry W, Jameson J, Cruz J, Ennis FA, Rothman AL (1999) Partial agonist
    effect influences the CTL response to a heterologous dengue virus serotype. J Immunol
    163:2754–2760

HLA and Other Gene Associations with Dengue Disease Severity

H.A.F. Stephens

Contents

Abstract Large case control gene association studies have been performed on cohorts of dengue virus (DENV) infected patients identified in mainland Southeast Asia, South Asia and the Caribbean. Candidate genes that have shown statistically significant associations with DENV disease severity encode HLA molecules, cell receptors for IgG (FcGII), vitamin D and ICAM3 (DCSIGN or CD209), pathogen recognition molecules such as mannose binding lectin (MBL), blood related antigens including ABO and human platelet antigens (HPA1 and HPA2). In ethnic Thais with secondary infections a variety of HLA class I alleles (HLA-A*0203, *0207, *A11, B*15, B*44, B*46, B*48, B*51, B*52), DCSIGN promoter polymorphisms and the AB blood group, independently associate with either susceptibility or resistance to dengue fever (DF) and the more severe dengue hemorrhagic fever (DHF). There is also evidence that some HLA associations with disease severity correlate with the DENV serotype inducing secondary infections. Taken together, there is now evidence that allelic

H.A.F. Stephens

Centre for Nephrology & The Anthony Nolan Trust, University College London, The Royal Free Hospital Campus, Rowland Hill Street, London, NW3 2PF, UK
e-mail: h.stephens@ucl.ac.uk

A.L. Rothman (ed.), *Dengue Virus*, Current Topics in Microbiology and Immunology 338,   99
DOI 10.1007/978-3-642-02215-9_8, © Springer-Verlag Berlin Heidelberg 2010

variants of multiple gene loci involved in both acquired and innate immune responses contribute significantly to DENV disease outcome and severity. Further analysis of the genetic basis of severe DENV disease in different at risk populations may contribute to the development of new preventative and therapeutic interventions.

1   Introduction

Exposure to the *flavivirus* dengue virus (DENV) invokes a variety of genetically-controlled immunological responses. These include the activation of T, B and natural killer (NK) cells, as well as the production of antibodies and a range of cytokines, which together can be either protective or detrimental to individuals exposed to DENV (Rothman 2004). There are four distinct DENV serotypes (DENV-1, -2, -3 and -4) that share 65–70% sequence homology (Green and Rothman 2006). Clinical symptoms range from the relatively benign dengue fever (DF), to the more severe dengue hemorrhagic fever (DHF) and dengue shock syndrome (DSS) (World Health Organisation, WHO, 1997). However, a significant proportion of DENV infections are subclinical (Endy et al. 2002), which may be due to exposure to less pathogenic DENV strains (Leitmeyer et al. 1999, Watts et al. 1999), or a genetically determined ability of the host to clear the virus before clinical symptoms can arise. Naturally acquired serotype-specific protective immunity is a well recognised outcome of DENV infection (Rothman 2004). By contrast, cross-reactive partial immunity to previously unseen DENV serotypes contributes to severe disease in some individuals undergoing secondary infections (Green and Rothman 2006). Thus, a complex interplay between genetic variants of DENV and the host immune response is likely to determine the outcome of primary and secondary infections.

Genetic variants of DENV serotypes can differ in the severity of the clinical infection they induce (Watts et al. 1999). There is also evidence that severe DHF is a rare consequence of DENV exposure in Africans (Halstead et al. 2001, Sierra et al. 2007a), which raises the possibility that human resistance genes and natural immunity exist in certain ethnic groups. Genetic screening of populations for associations with infectious diseases has recently become attainable with the availability of high resolution human genome sequence, generated by high-throughput molecular typing. Two major approaches are widely used. These are candidate gene analyses that assume likely genetic targets worthy of detailed investigation, or blind association studies where the genome of patients is screened with a large number of high resolution markers in a linkage analysis for susceptibility genes (Tibayrenc 2007). At present the candidate gene approach has been applied to cohorts of dengue patients and the most scrutinised gene loci are those encoding human leukocyte antigens (HLA).

1.1 HLA Genes: Function, Polymorphism and Relevance to Dengue

Located on the short arm of human chromosome 6 lies a dense cluster of some 400 genes called the major histocompatibility complex or MHC. Nearly a third of the expressed MHC loci are functionally involved in both acquired and innate immune responses to microbes (Horton et al. 2004). These include genes encoding classical HLA class I and II molecules, whose biological function is to present short microbial-derived peptides to the antigen-specific T cell receptors of CD8+ cytolytic and CD4+ helper cells, respectively. HLA class I molecules are also recognised by another group of receptors that determine whether cells with innate NK activity spontaneously lyse virally-infected cells (Parham, 2005a). Thus, HLA molecules are essential recognition elements of acquired and innate immune responses to viruses.

The genes encoding classical HLA class I (HLA-A, -B, -C) and class II (HLA-DR, DQ and DP) are the most polymorphic in the human genome. Each locus encodes hundreds of alleles (EBI/EMBL 2008), which indicates that HLA loci are still evolving probably under selective pressure exerted by microbes (Parham and Ohta 1996, Prugnolle et al. 2005). At the population level HLA loci show considerable differences in allele frequencies both within and between different ethnic groups (Chandanayingyong et al. 1997, Cao et al. 2004, Middleton et al. 2007). Another feature of HLA genes is the phenomenon of linkage disequilibrium, where certain alleles encoded by different loci tend to form stable combinations or haplotypes. The composition and frequency of HLA haplotypes also varies considerably between different ethnic groups (Chandanayingyong et al. 1997, Cao et al. 2004, Middleton et al. 2007).

The extreme polymorphism of HLA class I and II genes has a profound influence on the biological function of their products. Most polymorphism results in significant amino-acid changes in the first and second extra-cellular protein domains of HLA molecules, which affect the binding and subsequent presentation of microbial peptides to antigen-specific T cells (Parham 2005b). Allelic variants of HLA-A, -B, -DR, -DQ and -DP, present DENV-derived peptides to antigen-specific CD8+ and CD4+ T cells (Bashyam et al. 2006, Zivna et al. 2002, Mathew et al. 1998, Kurane et al. 1991, Green et al. 1997). DENV also up-regulates HLA expression in infected cells (Libraty et al. 2001). Thus, HLA class I and II molecules clearly play an important role in the host immune response to DENV, which renders them an obvious target for genetic association studies.

1.2 HLA Gene Association Studies with Dengue

The interpretation of case-control HLA disease association studies can be complicated by a variety of factors (Cardon and Bell 2001, NCI-NHGRI, National Cancer Institute-National Human Genome Research Institute, 2007): size and ethnicity of the target populations, frequency and composition of the detected

HLA alleles and haplotypes, degree of resolution of typing methods and clinical or biological definitions of the disease outcomes under investigation. Large cohorts of several hundred patients and controls are essential to provide sufficient statistical power to detect genuine non-random HLA associations with disease phenotype. Adequately powered HLA association studies with DENV infections have been performed in target populations located in mainland S.E. Asia, S. Asia and the Caribbean, in addition to other smaller studies performed in Central and S. America (Table 1).

1.3 HLA and Dengue in Southeast and South Asia

In mainland S.E. Asia all four DENV serotypes are in seasonal circulation. Recent epidemiologic analysis suggests that DENV is amplifying in the large cosmopolitan cities of the region such as Bangkok (Cummings et al. 2004), where ethnic diversity and population density are at their greatest. Here immunological pressure probably induces positive selection on DENV (Twiddy et al. 2002), before the fittest serotypic strains transmit out in cyclical waves across the country and into adjacent populations (Cummings et al. 2004).

The populations of SE Asia that have been targeted for genetic association studies with DENV are the ethnic Thais and Vietnamese Kinh (Table 1). Taken together, these populations comprise some 140 million people but share relatively similar HLA allele profiles (Chandanayingyong et al. 1997, Stephens et al. 2000, Hoa et al. 2008). An analysis of HLA-A and –B class I allele profiles in a carefully defined cohort of Thai DF and DHF patients with either primary or secondary infections of known DENV serotype (Stephens et al. 2002), has in part confirmed previously reported associations with DHF in ethnic Thais (Chiewsilp et al. 1981) and Vietnamese (Loke et al. 2001). So far, HLA class I associations with disease severity have only been detected in Thai patients undergoing secondary infections and thereby immunologically primed by previous exposure to DENV (Table 1). This would imply that HLA class I-restricted cross-reactive T cell driven immune responses are contributing to the pathology of DENV. By contrast, HLA class I associations have been also identified with primary infections in the Vietnamese Kinh population (Lan et al. 2008). Some HLA associations with disease severity in ethnic Thais relate to the infecting DENV serotype, with DENV-3-associated HLA allele profiles differing in part from those associated with DENV-1 and -2 (Table 1). Similarly, DENV-2-specific HLA associations have also been identified in Vietnamese patients (Lan et al. 2008). Furthermore, associations detected between HLA-B*51 and DHF and the related B*52 allele with less severe DF in SE Asians (Stephens et al. 2002), are strikingly analogous to those identified in Behcets disease in multiple populations, where HLA-B*51 (but not HLA-B*52) associates strongly with this systemic vasculitis of the mucosa (Verity et al. 2003). The detection of an equivalent genetic correlation between secondary DHF and a disease of unknown aetiology such as Behcets, which is thought to be induced or

Table 1 HLA and other MHC encoded gene associations with dengue fever (DF) and dengue hemorrhagic fever (DHF) in different ethnic groups

Ethnic Group / location (size of study)	Positive association (Susceptibility)		Negative association (Resistance)		Reference
	DF	DHF	DF	DHF	
ASIA					
Thais / SE Asia (n=87)		HLA-A2[a]		HLA-B13	Chiew-silp et al. (1981)
Thais / SE Asia (n=263)	Secondary Infection: HLA-A*0203[a] (all DENV serotypes)	Secondary Infection: HLA-A*0207[a] (DENV1 and DENV2 only) HLA-A*11 (DENV3 only)	Secondary Infection:	Secondary Infection: HLA-A*33 (trend only, all DENV serotypes)	Stephens et al. (2002)
	HLA-B*51 (DENV3 only)	HLA-B*51 (all DENV serotypes)	HLA-B*15[b] (B62, B76 and B77, all DENV serotypes)	HLA-B*15[b] (B76 and B77, all DENV serotypes)	
	HLA-B*52 (DENV1 and DENV2 only)	HLA-B*46 (DENV1 and DENV2 only)		HLA-B*44 (all DENV serotypes)	
Thais / SE Asia (n=435)		Secondary Infection: HLA-B*48 (all DENV serotypes) TNF (tumour necrosis factor, pos -238)			Vejbaeysa et al. (2009)
Vietnamese / SE Asia (n=352)		HLA-A24 (DHF grades 3 and 4)		HLA-A33 (DHF grades 3 and 4)	Loke et al. (2001)
Vietnamese / SE Asia (n=629)		Primary infection: HLA*24 (alleles with histidine at pos 70 in DSS only)		Secondary infection: HLA-DRB1*0901 (DSS, DENV2 only)	Lan et al. (2008)
SW Indians / S Asia (n=197)		Primary infection: TAP1 (transporter associated		Primary infection: TAP1 (pos379)	Soundravally and Hoti. (2007, 2008)

(continued)

Table 1 (continued)

Ethnic Group / location (size of study)	Positive association (Susceptibility)		Negative association (Resistance)		Reference
	DF	DHF	DF	DHF	
ASIA		with antigen processing, pos 333			
AMERICAS					
Cubans / Caribbean (n=82)		HLA-A1		HLA-A29	Paradoa Perez et al. (1987)
Cubans / Caribbean (n=120)	HLA-B*15 (DENV2)	Secondary Infection: HLA-A*31 (DENV2) HLA-B*15 (DENV2)	HLA-DRB1*04 (DENV2) HLA-DRB1*07 (DENV2)	Secondary Infection: HLA-DRB1*04 (DENV2) HLA-DRB1*07 (DENV2)	Sierra et al. (2007)
Mexicans / C America (n=47)			HLA-DRB1*11	HLA-DRB1*04	La Fleur et al. (2002)
Venezuelans / S America (n=66)		TNF (pos -308)			Fernandez-Mestre et al. (2004)
Brazilians / S America (n=64)	Primary Infection HLA-DR1, DQ1 (DENV1 only)				Polizel et al. (2004)

[a] Serologically defined HLA alleles are given without an asterisk, e.g., HLA-A1. Molecularly defined alleles are defined with an asterisk, e.g., HLA-A2. Molecularly defined alleles are defined with an asterisk, e.g., HLA-A*0203 or A*0207 (allelic variants of HLA-A2)

[b] The molecularly defined HLA-B*15 group of related alleles, includes the serologically defined HLA-B62, -B76 and -B77 alleles

exacerbated by exposure to microbial infection (Verity et al. 2003), raises the possibility that some of the severe manifestations of DENV infection are of an autoimmune nature and driven by HLA-restricted responses.

Tumour necrosis factor (TNF) and lymphotoxin alpha (LTA) are important vasocactive modulators of the immune response and are known to be up-regulated in DHF infections (Green et al. 1999). TNF and LTA are encoded by adjacent gene loci in the central or class III region of the MHC, in-between HLA class I and II genes (Horton et al. 2004). The analysis of TNF and LTA genes in ethnic Thais has recently revealed specific combinations of TNF, LTA and HLA class I alleles that associate with secondary DHF in ethnic Thais and correlate with in vivo intra-cellular production of LTA and TNF during the acute viraemic phase of infection (Vejbaesya et al. 2009). Similarly, case-control studies in S. Asians of the MHC-encoded transporters associated with antigen processing (TAP) genes, which are involved in loading of antigenic peptides into HLA molecules, have also shown associations with DHF (Soudravally and Hoti 2007, 2008). However, defining a role for TAP polymorphism in susceptibility to infections is complicated by these genes being in linkage disequilibrium with HLA class II genes (Carrington et al. 1993, Klitz et al. 1995). In the future, suitably powered studies incorporating stratification for linkage disequilibrium with class II genes, may resolve whether genetic variants of TAP genes play a primary role in DENV infections.

1.4 HLA and Dengue in the Caribbean, Central and South America

In this region intermittent transmission of specific DENV serotypes occurs in relatively heterogeneous insular and mainland populations, composed of admix-tures of African, European Caucasoid, Asian and indigenous Amerindian decent. The genetic diversity of these at risk populations is clearly seen in the HLA allele and haplotype profiles reported in this region (Ferrer et al. 2007, Middleton et al. 2007). To date, the largest HLA association study performed in this region has been in patients identified in Cuba (Table 1), where an insular population of predomi-nantly European and African heritage was exposed to DENV-2 in a localised outbreak in 1981, after previous immunologic priming by an earlier exposure to DENV-1 (Sierra et al. 2007b). Both HLA class I and II genes were retrospectively analysed in this study and revealed similar associations with both DF and DHF patients when compared to a control group (Table 1). HLA-B*15, which represents a large group of diverse alleles (EBI/EMBL 2008), was significantly increased in frequency in both DF and DHF when compared to controls in the Cuban study (Sierra et al. 2007b). However, the frequency of HLA-B*15 does vary considerably between the various ethnic admixtures of the Cuban population and the control frequency in the Sierra et al. study (2007b) was considerably lower than other reports for the Cuban population (Ferrer et al. 2007). Similarly, the HLA class II associations with clinical DENV infection reported in Cuba, HLA-DRB1*04 and

*07 (Table 1), are of interest as they confirm in part a smaller study in Mexicans (La Fleur et al. 2002). Nevertheless, a relatively large proportion (23.5%) of HLA-DRB1 alleles in the Cuban patients was undefined (Sierra et al. 2007b), which, if resolved, could affect the overall significance of the HLA class II associations with DENV infections reported in this study.

Given that the "gold standard" for establishing genetic associations with any disease is to replicate statistical significance in multiple groups of patients and controls (Cardon and Bell 2001, NCI-NHGRI 2007), the HLA associations in Cubans (Sierra et al. 2007b) should be tested again with new DENV cohorts recruited in the same location or other populations in this region. The same strategy applies to the HLA associations reported in some of the smaller case control studies performed in Central and South America (Table 1). For example, a particular polymorphism at position −308 of the TNF gene, has been reported to be significantly increased in just 25 Venezuelan DHF patients (Fernadez-Mestre et al. 2004). Yet when the same polymorphism was analysed in much larger cohorts of Vietnamese and Thai DHF patients (n = 352 and 435, respectively), no association with disease phenotype was observed (Loke et al. 2001, Vejbaesya et al. 2009), despite the TNF-308 polymorphism being present in these populations and being associated with other infections such as typhoid fever (Dunstan et al. 2001). Moreover, given that the TNF-308A/G polymorphism is known to be in strong linkage disequilibrium with certain HLA-DRB1 alleles (Dunstan et al. 2001), the analysis of HLA class II allele profiles in the Venezuelan DF and DHF cohort could prove to be informative.

2   Other Non-MHC Gene Associations with Dengue

Large case-control studies examining a variety of single-nucleotide polymorphisms (SNPs) in a variety of genes encoded outside of the MHC have been performed in several hundred DENV patients recruited in Vietnam, Thailand, India and Brazil (Table 2). Selected target genes that have shown statistically significant associations with the severity of DENV infections encode cell receptors, mediators of immune activation, pathogen recognition molecules and blood-related antigens. Polymorphisms that affect the function of Vitamin D receptors (VDR), an immune mediator and the receptor for the Fc domain of IgG2 (FcGRIIA or CD31), which facilitates antibody-dependent enhancement of DENV infection (Littua et al. 1990), show moderate associations with resistance to the most severe form of DHF (WHO grades 3 and 4) in Vietnam (Loke et al. 2002). However, a putative SNP-defined promoter polymorphism in the gene encoding dendritic cell-specific ICAM-3 grabbing non-integrin (DCSIGN or CD209), which can act as a receptor for DENV (Tassaneetrithep et al. 2003), has repeatedly been shown to be protective against developing DF but not DHF in ethnic Thais (Sakuntabhai et al. 2005). This provides further evidence of independent genetic risk factors that associate with either DF or DHF in mainland SE Asian populations (Tables 1 and 2). Considerable genetic diversity of DCSIGN has also been reported in different ethnic groups, with

Table 2 Non-MHC encoded gene associations with dengue fever (DF) and dengue hemorrhagic fever (DHF) in different ethnic groups

Ethnic Group / location (size of study)	Positive association (Susceptibility)		Negative association (Resistance)		Reference
	DF	DHF	DF	DHF	
ASIA					
Vietnamese / SE Asia (n=400)				Vitamin D receptor (DHF grades 3 and 4) Fc Gamma receptor IIA / CD32 (DHF grades 3 and 4)	Loke et al. (2002)
Thais / SE Asia (n=606)			CD209 / DC-SIGN1 promoter		Sakuntabhai et al. (2005)
Thais / SE Asia (n=399)		Secondary Infection: Blood group AB (DHF grade 3 only)			Kalayanarooj et al. (2007)
South Indians / S Asia (n=197)	Human Platelet Antigen 1 (HPA1)	Human Platelet Antigen 2 (HPA2)			Soundravally and Hoti, (2007)
AMERICAS					
Venezuelans / S America (n=66)		IL10 (pos-1082)			Fernandez-Mestre et al. (2004)
Brazilians / S America (n=110)	Mannose binding lectin (MBL pos 52, 54, 57)				Acioli-Santos et al. (2008)

African populations being the most polymorphic (Boily-Larouche et al. 2007). This diversity may relate to higher local pathogen diversity and pressure on African populations and would be analogous to the probable evolutionary forces driving HLA polymorphism (Prugnolle et al. 2005). Furthermore, in a recent study performed in Brazil, a combination of SNP markers associated with the low producer phenotype of mannose binding lectin (MBL), which is a complement-activating pathogen recognition molecule, significantly associate with DF and not DHF (Acioli-Santos et al. 2008).

One of the characteristic clinical features of DHF and DSS is the phenomenon of vascular permeability, plasma leakage and the development of thrombocytopenia (Rothman 2004). Not surprisingly a variety of blood related antigens have come under the scrutiny of genetic epidemiologists (Table 2). For example, a preponderance of blood group AB has been reported in ethnic Thai DHF patients with secondary infections resulting in the most severe disease (WHO grades 3 and 4) (Kalayanarooj et al. 2007). This phenotype is characterised by the absence of antibodies to immunodominant blood group carbohydrate antigens. Such antibodies may also cross-react with glycosylated DENV proteins and thereby provide some degree of protection in DENV-exposed individuals. Furthermore, as antibody-dependent binding of DENV-2 to platelets has been shown to occur in the absence of functional FcγR (Wang et al. 1995), other human platelet antigen (HPA) polymorphisms have now been screened for disease associations with dengue (Table 2). In a recent study conducted in India SNPs identifying structural variants of HPA1 and HPA2, which facilitate platelet interaction with the endovascular wall and von Willebrand factor, appear to associate with susceptibility to either DF or DHF, respectively (Soundravelly and Hoti 2007).

Not all population-based investigations have revealed significant genetic associations with DENV infections. Neutral associations have been reported for IL-4 promoters, IL-1 receptor antagonist, interferon-gamma, IL-6, TGF-beta and CD45 (Loke et al. 2002, Fernandez-Mestre et al. 2004, Ward et al. 2006). Nevertheless, the picture that has emerged from the analysis of the non-MHC association studies is that variability in genes involved with the innate host immune response to DENV are associated with disease severity (Table 2).

3 Host Gene Expression Studies

The expression of mRNA transcripts and protein products of select immune response genes has been measured in the plasma and peripheral blood of DF and DHF patients during the acute and convalescent phases of DENV infection. High levels of soluble IL-2 receptor, soluble CD8, soluble TNF receptor, IL-10, macrophage inhibitory factor and TNF all correlate with severity of DENV disease and indicate a high degree of T cell activation (Green et al. 1999, Gagnon et al. 2002, Chen et al. 2006). With the advent of microarray technology, the expression levels of a vast range of host genes has been screened in cells infected with DENV in vitro

(Moreno-Altamirano et al. 2004, Fink et al. 2007) and in the peripheral blood of DF and DHF patients (Simmons et al. 2007, Fink et al. 2007). The sheer output of information on gene transcription generated by this technology is formidable and will require laborious confirmation at the protein or phenotype level, with careful correction for confounding variables such as host variability and post-transcriptional modification events. Nevertheless, preliminary studies have indicated that a diverse set of genes associated with B cell activation, the endoplasmic reticulum, apoptosis, oxygen transport and the innate immune response are expressed in abundance in Vietnamese patients with DSS (Simmons et al. 2007). However, an alternative and more direct functional genomics approach to identify candidate genes essential for DENV replication has been the use of a genome-wide RNA interference screen on DENV-infected cells. Here libraries of small interfering RNAs (siRNAs) are used to selectively block the function of thousands of human genes and by a process of elimination identify those genes that are essential for DENV replication and therefore potential therapeutic targets (Krishnan et al. 2008).

4   New Vistas Along the Genetic Highway

The scale, resolution and throughput of molecular typing of complex and highly polymorphic candidate genes such as HLA has now reached a point where immunogenetic surveillance within and between populations can be performed at a level that will achieve considerable statistical power in disease association studies. New microsphere-based methods enable host and pathogen genetic polymorphism, as well as serum levels of cytokines and antibodies, to be measured using the same technical platform, which can be easily transferred to field study sites. The throughput of large gene association studies could be further expedited by establishing well-coordinated collaborations between laboratories working with DENV-exposed populations in different regions, through the sharing of genetic typing materials and methods, as well as implementing consensus and robust methods of statistical analyses.

Information derived from the proteomic analysis of small antigenic peptides bound to HLA molecules (Parham 2005b) has been used in the rational selection of target antigens in modern vaccines (Flower 2003, Nussbaum et al. 2003). This approach has recently been applied to screen DENV proteins for conserved HLA-binding immunogenic peptides with the aim of designing new and effective vaccines (Khan et al. 2008). Purified crystals of an HLA class I molecule (HLA-A*1101) bound to a variety of immunogenic DENV-derived peptides have also been produced (Chotiyarnwong et al. 2007). With these materials available it should be possible to define how potentially cross-reactive DENV peptides are presented to T cells by HLA-A*1101, which could be of relevance to DENV vaccine design and testing in mainland SE Asian populations, in which HLA-A*1101 is relatively common (Chandanyingyong et al. 1997). Similarly, antigen-loaded HLA class I tetramers labelled with fluorescent tags have been informative in real-time flow cytometric analysis of the level, specificity, persistence and recall of CTL responses to DENV (Mongkolsapaya et al. 2003, Dong et al. 2007).

The effect of NK cells and the genetic variability of their numerous receptors that recognise HLA class I molecules (Parham 2005a), have been the subject of investigations in other flavivirus infections such as Hepatitis C (Khakoo et al. 2004). An equivalent analysis of NK receptor gene profiles in relation to DENV infection outcome may be worth performing in the large cohorts of patients that have already been HLA typed in SE Asia and the Caribbean, particularly as considerable information is already available on NK receptor gene profiles in some of these populations (Norman et al. 2001, Norman et al. 2004, Norman et al. 2007). The genetic basis for the relatively rare incidence of DHF in DENV-exposed African populations (Halstead et al. 2001, Sierra et al. 2007a) has yet to be explored at the molecular level. Such studies may prove highly informative and reveal evolutionary effects exerted by other pathogenic flaviviruses such as yellow fever, which has an equivalent route of transmission to DENV. Likewise, a comprehensive genetic analysis has yet to be performed on individuals with documented sub-clinical or asymptomatic DENV infections in DHF-susceptible populations with endemic transmission. In many respects the DENV field is entering a potentially new phase in terms of genetic studies. With the methods and materials now available, many basic questions regarding susceptibility and resistance to severe disease can be addressed at the molecular level, which may contribute to the development of new DENV-specific preventative and therapeutic interventions.

Acknowledgments This work was supported by a Program Project grant from the National Institutes of Health USA (NIH-P01AI34533)

References

Acioli-Santos B, Segat L, Dhalia R, Brito CAA, Braga-Neto UM, Marques ETA, Crovella S (2008) MBL2 gene polymorphisms protect against development of thrombocytopenia associated with severe dengue phenotype. Hum Immunol 69(2):122–128

Bashyam HS, Green S, Rothman AL (2006) Dengue virus-reactive CD8+ T cells display quantitative and qualitative differences in their response to variant epitopes of heterologous viral serotypes. J Immunol 176:2817–2824

Boily-Larouche G, Zijenah LS, Mbizvo M, Ward BJ, Roger M (2007) DC-SIGN and DC-SIGNR genetic diversity among different ethnic populations: potential implications for pathogen recognition and disease susceptibility. Hum Immunol 68:523–530

Cao K, Moormann AM, Lyke KE, Masaberg C, Sumba C, Doumbo OK, Koech D, Lancaster A, Nelson M, Meyer D, Single R, Hartzman RJ, Plowe CV, Kazura J, Mann DL, Sztein MB, Thomson G, Fernandez-Vina MA (2004) Differentiation between African populations is evidenced by the diversity of alleles and haplotypes of HLA class I loci. Tissue Antigens 63:293–325

Cardon LR, Bell JI (2001) Association study designs for complex diseases. Nature Rev Genetics 2:91–99

Carrington M, Colonna M, Spies T, Stephens JC, Mann DL (1993) Haplotypic variation of transporter associated with antigen processing (TAP) genes and their extension of HLA class II region haplotypes. Immunogenetics 37:266–273

Chandanayingyong D, Stephens HAF, Klaythong R, Sirkong M, Udee S, Longta P, Chantangpol R, Bejrachandra S, Runruang E (1997) HLA-A, -B, -DRB1, -DQA1 and -DQB1 polymorphism in Thais. Hum Immunol 53:174–182

Chen LC, Lei HY, Liu CC, Shiesh SC, Chen SH, Liu HS, Lin YS, Wang ST, Shyu HW, Yeh TM (2006) Correlation of serum levels of macrophage inhibitory factor with disease severity and clinical outcome in dengue patients. Am J Trop Med Hyg 74:142–147

Chiewsilp P, Scott RM, Bhamarapravati N (1981) Histocompatibility antigens and dengue hemorrhagic fever. Am J Trop Med Hyg 30:1100–5

Chotiyarnwong P, Stewart-Jones GB, Tarry MJ, Dejnirattisai W, Siebold C, Koch M, Stuart DI, Harlos K, Malasit P, Screaton G, Mongkolsapaya J, Jones EY (2007) Humidity control as a strategy for lattice optimization applied to crystals of HLA-A*1101 complexed with variant peptides from dengue virus. Acta Crystallogr Sect F Struct Biol Cryst Commun 63:386–392

Cummings DAT, Irizarry RA, Huang NE, Endy TP, Nisalak A, Ungchusak K, Burke DS (2004) Travelling waves in the occurrence of dengue hemorrhagic fever in Thailand. Nature 427:344–347

Dong T, Moran E, Chau NV, Simmons C, Luhn K, Peng Y, Wills B, Dung NP, Thao LTT, Hian TT, McMichael A, Farrar J, Rowland-Jones S (2007) High pro-inflammatory cytokine secretion and loss of high avidity cross-reactive cytotoxic T-cells during the course of secondary dengue virus infection. PLoS ONE 2(12):e1192

Dunstan SJ, Stephens HAF, Blackwell JM, Duc CM, Lanh MN, Dudbridge F, Phuong CXT, Luxemberger C, Wain J, Ho VA, Hien TT, Farrar J, Dougan G (2001) Genes of the Class II and Class III Major Histocompatibility Complex are associated with Typhoid Fever in Vietnam. J Inf Dis 183:261–268

EBI/EMBL. (2008) IMGT/HLA Database, release 2.20.0. European Bioinformatics Institute/ European Molecular Biology Laboratory. Available at www.ebi.ac.uk/imgt/hla

Endy TP, Chunsuttiwat S, Nisalak A, Nisalak A, Libraty DH, Green S, Rothman AL, Vaughan DW, Ennis FA (2002) Epidemiology of inapparent and symptomatic dengue virus infection: a prospective study of primary school children in Kamphaeng Phet, Thailand. Am J Epidemiol 156:40–51

Fernandez-Mestre MT, Gendzekhadze K, Rivas-Vetencourt P, Layrisse (2004) TNF-alpha-308A allele, a possible severity risk factor of hemorrhagic manifestation in dengue fever patients. Tissue Antigens 64:469–472

Ferrer A, Nazabal M, Companioni O, Fernandez de Cossio ME, Camacho H, Cintado A, Benitez J, Casalvilla R, Sautie M, Villareal A, Diaz T, Marrero A, Fernandez de Cossio J, Hodelin A, Leal L, Ballester L, Novoa LI, Middleton D, Duenas M (2007) HLA class I polymorphism in the Cuban population. Hum Immunol 68:918–927

Fink J, Gu F, Ling L, Tolfvenstam T, Olfat F, Chin KC, Aw P, George J, Kuznetsov VA, Schreiber M, Vasudevan SG, Hibberd ML (2007) Host gene expression profiling of dengue virus infection in cell lines and patients. PLoS Negl Trop Dis 1(2):e86

Flower DR (2003) Towards in silico prediction of immunogenic epitopes. Trends Immunol 24:667–674

Gagnon SJ, Mori M, Kurane I, Green S, Vaughan DW, Kalayanarooj S, Suntayakorn S, Ennis FA, Rothman AL (2002) Cytokine gene expression and protein production in peripheral blood mononuclear cells of children with acute dengue virus infections. J Med Virol 67:41–46

Green S, Kurane I, Pincus S, Paoletti E, Ennis FA (1997) Recognition of dengue virus NS1-NS2a proteins by human CD4+ cytotoxic T lymphocyte clones. Virology 234:383–386

Green S, Vaughn DW, Kalayanarooj S, Nimmannitya S, Suntayakorn S, Nisalak A, Lew R, Innis BL, Kurane IRAL, Ennis FA (1999) Early immune activation in acute dengue is related to development of plasma leakage and disease severity. J Infect Dis 179:755–762

Green S, Rothman A (2006) Immunopathogic mechanisms in dengue and dengue hemorrhagic fever. Curr Opin Inf Dis 19:429–436

Halstead SB, Streit TG, Lafontant JG, Putvantana R, Rusell K, Sun W, Kanesa-Thasan N, Hayeas CG, Watts DM (2001) Haiti: absence of dengue hemorrhagic fever despite dengue virus transmission. Am J Trop Med Hyg 65:180–183

Hoa BK, Hang NTL, Kashiwase K, Ohashi J, Lien LT, Shojima J, Hijikata M, Sakurada S, Satake M, Tokunaga K, Sasazuki T, Keicho N (2008) HLA-A, -B, -C, -DRB1 and -DQB1 alleles and haplotypes in the Kinh population in Vietnam. Tissue Antigens 71:127–134

Horton R, Wilming L, Rand V, Lovering RC, Bruford EA, Khodiyar VK, Lush MJ, Povey S, Talbot CC, Wright MW, Wain HM, Trowsdale J, Ziegler A, Beck S (2004) Gene map of the extended human MHC. Nature Rev Immunol 5:889–899

Kalayanarooj S, Gibbons RV, Vaughan D, Green S, Nisalak A, Jarman RG, Mammen MP, Perng GC (2007) Blood group AB is associated with increased risk for severe dengue disease in secondary infections. J Inf Dis 195:1014–1017

Khakoo SI, Thio CL, Martin MP, Brooks CR, Gao X, Astemborski J, Cheng J, Goedert JJ, Vlahov D, Hilgartner M, Cox S, Little AM, Alexander GJ, Cramp ME, O'Brien SJ, Rosenberg WM, Thomas DL, Carrington M (2004) HLA and NK inhibitory receptor genes in resolving hepatitis C virus infection. Science 305:872–874

Khan AM, Miotto O, Nascimento EJM, Srinivasan KN, Heiny AT, Zan GL, Marques ET, Tan TW, Brusic V, Salmon J, August JT (2008) Conservation and variability of dengue virus proteins: implications for vaccine design. PLoS Neg Trop Dis 2(8):e272

Klitz W, Stephens JC, Grote M, Carrington M (1995) Discordant patterns of linkage disequilibrium of the peptide-transporter loci within the HLA class II region. Am J Hum Genet 57:1436–1444

Krishnan MN, Ng A, Sukumaran B, Gilfoy FD, Uchil PD, Sultana H, Brass AL, Adametz R, Tsui M, Qian F, Montgomery RR, Lev S, Mason PW, Koski RA, Elledge SJ, Xavier RJ, Agaisse H, Fikrig E (2008) RNA interference screen for human genes associated with West Nile virus infection. Nature 455:242–245

Kurane I, Brinton MA, Samson AL, Ennis FA (1991) Dengue-specific, human CD4+ CD8-cytotoxic T-cell clones: multiple patterns of virus cross-reactivity recognized by NS3-specific T-cell clones. J Virol 65:1823–1828

La Fleur C, Granados J, Vargas-Alarcon C, Higuera L, Hernandez-Pacheo G, Cutino-Moguel T, Rangel H, Figuera R, Acosta M, Lazcano E, Ramos C (2002) HLA-DR antigen frequencies in Mexican patients with dengue virus infection: HLA-DR4 as a possible genetic resistance factor for dengue hemorrhagic fever. Hum Immunol 63:1039–1044

Lan NTP, Kikuchi M, Huong VTQ, Ha DQ, Thuy TT, Tham VD, Tuan VV, Nga CTP, Dat TV, Oyama T, Morita K, Yasunami M, Hirayama K (2008) Protective and enhancing HLA alleles, HLA-DRB1*0901 and HLA-A*24, for severe forms of dengue virus infection, dengue hemorrhagic fever and dengue shock syndrome. PLoS Neg Trop Dis 2(10):e304

Leitmeyer KC, Vaughan DW, Watts DM, Salas R, Villalobos de Chacon I, Ramos C, Rico-Hesse R (1999) Dengue virus structural differences that correlate with pathogenesis. J Virol 73:4738–4747

Libraty DH, Pichyangkul S, Ajariyakhajorn C, Endy TP, Ennis FA (2001) Human dendritic cells are activated by dengue virus infection: enhancement by gamma interferon and implications for disease pathogenesis. J Virol 75:3501–3508

Littua R, Kurane I, Ennis FA (1990) Human IgG Fc receptor II mediates antibody-dependent enhancement of dengue virus infection. J Immunol 141:3183–3186

Loke H, Bethell DB, Phuong CXT, Dung M, Schneider J, White NJ, Day NP, Farrar J, Hill AVS (2001) Strong HLA class I-restricted T cell responses in dengue hemorrhagic fever: a double-edged sword? J Inf Dis 184:1369–73

Loke H, Bethell D, Phuong CXT, Day N, White N, Farrar J, Hill A (2002) Susceptibility to dengue hemorrhagic fever in Vietnam: evidence of an association with variation in the vitamin D receptor and FC gamma receptor IIA genes. Am J Trop Med Hyg 67:102–106

Mathew A, Kurane I, Green S, Stephens HAF, Vaughn DW, Kalayanarooj S, Suntayakorn S, Chandanayingyong D, Ennis FA, Rothman AL (1998) Predominance of HLA-restricted cytotoxic T-lymphocyte responses to serotype- cross-reactive epitopes on non-structural proteins following natural secondary dengue virus infection. J Virol 72:3999–4004

Middleton, D., Menchaca, L., Rood, H. and Komerofsky, R. (2007) Allele frequencies in worldwide populations: New allele frequency database. Available at www.allelefrequencies.net

Mongkolsapaya J, Dejinirattisai W, Xu X, Vasanwathanana S, Tangthawornchaikul N, Chairunsri A, Sawasdivorn S, Duangchinda T, Dong T, Rowland-Jones S, Yenchitsomanus PT, McMichael A, Malasit P, Screaton G (2003) Original antigenic sin and apoptosis in the pathogenesis of dengue hemorrhagic fever. Nature Med 9:921–927

Moreno-Altamirano MM, Romano M, Legorreta-Herrera M, Sanchez-Garcia FJ and Colston MJ. (2004) Gene expression in human macrophages infected with dengue virus serotype-2. Scand J Immunol 60:631–638

NCI-NHGRI (National Cancer Institute-National Human Genome Research Institute) (2007) Replicating genotype-phenotype associations. Nature 447:655–660 Working Group on Replication in Association Studies, Chanock, S.J., Manolio T., Boehnke, M., Boerwinkle, E., Hunter, D.J., Thomas, G., Hirschhorn, J.N., Abeciasis, G., Altshuler, D., Bailey-Wilson, J.E., Brooks, L.D., Cardon, L.R., Daly, M., Donnelly, P., Fraumeni Jr, J.F., Freimer, N.B., Gerhard, D.S., Gunter, C., Guttmacher, A.E., Guyer, M.S., Harris, E.L., Hoh, J., Hoover, R., Kong, C.A., Merikangas, K.R., Morton, C.C., Palmer, L.J., Phimister, E.G., Rice, J.P., Roberts, J., Rotimi, C., Tucker, M.A., Vogan, K.J., Wacholder, S., Wijsman, E.M., Winn, D.M. and Collins, F.S

Norman PJ, Stephens HAF, Verity DH, Chandanayingyong D, Vaughan RW (2001) Distribution of natural killer cell immunoglobulin-like receptor sequences in three ethnic groups. Immunogenetics 52:195–205

Norman PJ, Cook MA, Carey BS, Carrington CVF, Verity DH, Hameed K, Ramdath RD, Chandanayingyong D, Leppert M, Stephens HAF, Vaughan RW (2004) SNP haplotypes and allele frequencies show evidence for disruptive and balancing selection in the human leukocyte receptor complex. Immunogenetics 56:225–237

Norman PJ, Abi-Rached L, Gendzekhadze K, Korbel D, Gleimer M, Rowley D, Bruno D, Carrington CVF, Chandanayingyong D, Chang Y-H, Crespi C, Saruhan-Direskeneli G, Fraser P, Hameed K, Kamkamidze G, Koram KA, Layrisse Z, Matamoros N, Mila J, Park MH, Pitchappan RM, Ramdath DD, Shiau M-Y, Stephens HAF, Struik S, Verity DH, Vaughan RW, Tyan D, Davis RW, Riley EM, Ronaghi M, Parham P (2007) Unusual selection on the KIR3DL1/S1 natural killer receptor in Africans. Nat Genet 39:1092–1099

Nussbaum AK, Kuttler C, Tenzer S, Hansjorg S (2003) Using the world wide web for predicting CTL epitopes. Curr Op Immunol 15:69–74

Paradoa Perez ML, Trujillo Y, Basanta P (1987) Association of dengue hemorrhagic fever with the HLA system. Haematologia 20:83–7

Parham P (2005a) MHC class I molecules and KIRS in human history, health and survival. Nature Rev Immunol 9:201–214

Parham P (2005b) Putting a face to MHC restriction. J Immunol 174:3–5

Parham P, Ohta T (1996) Population biology of antigen presentation by MHC class I molecules. Science 272:67–74

Polizel JR, Bueno D, Visentainer VEL, Sell AM, Borelli SD, Tsuento LT, Dalalio MMO, Coimbra MTM, Moliterno RA (2004) Association of human leukocyte antigen DQ1 and dengue fever in a white southern Brazilian population. Mem Inst Oswaldo Cruz 99:559–562

Prugnolle F, Manica A, Charpentier M, Guegan JF, Guernier V, Balloux F (2005) Pathogen-driven selection of worldwide HLA class I diversity. Curr Biol 15:1022–1027

Rothman AL (2004) Dengue: defining protective versus pathologic immunity. 2004. J Clin Invest 113:946–951

Sakuntabhai A, Turbpaiboon C, Casademont I, Chuansumrit A, Lowhnoo T, Kajaste-Rudnitski A, Kalayanarooj SM, Tangnararatchakit K, Tangthawornchaikul N, Vasanawathana S, Chaiyaratana W, Yenchitsomanus P, Suriyaphol P, Avirutnan P, Chokephaibulkit K, Matsuda F, Yoksan S, Jacob Y, Lathrop GM, Malasit P, Despres P, Julier C (2005)

A variant in the *CD209* promoter is associated with severity of dengue disease. Nat Genet 37:507–13

Sierra BC, Kouri G, Guzman MG (2007a) Race: a risk factor for dengue hemorrhagic fever. Arch Virol 152:533–542

Sierra B, Alegre R, Perez AB, Garcia G, Sturn-Ramirez K, Obansanjo O, Aguirre E, Alvarez M, Rodriguez-Roche R, Valdes L, Kanki P, Guzman MG (2007b) HLA-A, -B, -C, and -DRB1 allele frequencies in Cuban individuals with antecedents of dengue 2 disease: advantages of the Cuban population for HLA studies of dengue virus infection. Hum Immunol 68:531–540

Simmons CP, Popper S, Dolocek C, Chau TNB, Griffiths M, Dung NTP, Long TN, Hoang DM, Chau NV, Thao LTT, Hian TT, Relman DA, Farra J (2007) Patterns of host genome-wide transcription abundance in the peripheral blood of patients with acute dengue hemorrhagic fever. J Inf Dis 195:1097–1107

Soundravally R, Hoti SL (2007) Immunopathogenesis of dengue hemorrhagic fever and shock syndrome: role of TAP and HPA gene polymorphism. Hum Immunol 68:973–979

Soundravally R, Hoti SL (2008) Polymorphism of the TAP-1 and 2 gene may influence outcome of primary dengue viral infection. Scand J Immunol 67:618–625

Stephens HAF, Chandanayingyong D, Kunachiwa W, Sirikong M, Longta K, Maneemaroj R, Wongkuttiya D, Sittisombut N, Rungruang E (2000) A comparison of molecular HLA-DR and DQ allele profiles forming DR51-, DR52- and DR53-related haplotypes in five ethnic Thai populations from mainland Southeast Asia. Hum Immunol 61:1039–1047

Stephens HAF, Klaythong R, Sirikong M, Vaughn DW, Green S, Kalayanarooj S, Endy TP, Libraty DH, Nisalak A, Innis BL, Rothman AL, Ennis FA, Chandanayingyong D (2002) HLA-A and B allele associations with secondary dengue virus infections, correlate with disease severity and the infecting viral serotype in ethnic Thais. Tissue Antigens 60:309–318

Tassaneetrithep B, Burgess TH, Granelli-Peperno A, Trumpfheller C, Finke J, Sun W, Eller MA, Pattanapanyasat K, Sarasombath S, Birx DL, Steinman RM, Schlesinger S, Marovich MA (2003) DC-SIGN (CD209) mediates dengue virus infection of human dendritic cells. J Exp Med 197:823–829

Tibayrenc M (2007) Human genetic diversity and the epidemiology of parasitic diseases and other transmissible diseases. Adv Parasitol 64:377–428

Twiddy SS, Woelk CH, Holmes EC (2002) Phylogenetic evidence for adaptive evolution of dengue viruses in nature. J Gen Virol 83:1679–1689

Vejbaesya S, Luangtrakool P, Luangtrakool K, Kalayanarooj S, Vaughn DW, Endy TP, Mammen MP, Green S, Libraty DH, Ennis FA, Alan L, Rothman AL, Stephens HAF (2009) Tumor necrosis factor (TNF) and lymphotoxin-alpha (LTA) gene, allele, and extended HLA haplotype associations with severe dengue virus infection in ethnic Thais. J Inf Dis 199(10):1442–1448

Verity DH, Wallace GR, Vaughan RW, Stanford MR (2003) Behcets disease: from Hippocrates to the third millennium. Brit J Opthalmol 87:1175–1183

Wang S, He R, Patarapotikul J, Innis BL, Anderson R (1995) Antibody enhanced binding of dengue-2 virus to human platlets. Virology 213:254–257

Ward V, Hennig BJ, Hirai K, Tahara H, Tamori A, Dawes R, Saito M, Bangham C, Stephens H, Goldfeld A, Kunachiwa W, Leetrakool N, Hopkin J, Dunstan S, Hill A, Bodmer W, Beverley PCL, Tchilian EZ (2006) Geographical distribution and disease associations of the CD45 exon 6 138G variant. Immunogenetics 58:235–239

Watts DM, Porter KR, Putvatana P, Vasquez B, Calampa C, Hayes CG, Halsted SB (1999) Failure of secondary infection with American genotype dengue 2 to cause dengue hemorrhagic fever. Lancet 354:1431–1434

World Health Organisation (WHO) (1997) Dengue hemorrhagic fever. Diagnosis, treatment prevention and control, 2nd edn. WHO, Geneva, pp 1–84

Zivna I, Green S, Vaughn DW, Kalayanarooj S, Stephens HAF, Chandanayingyong D, Nisalak A, Ennis FA, Rothman AL (2002) T cell responses to an HLA-B*07-restricted epitope on the dengue NS3 protein correlate with disease severity. J Immunol 168:5959–5965

Vector Dynamics and Transmission of Dengue Virus: Implications for Dengue Surveillance and Prevention Strategies

Vector Dynamics and Dengue Prevention

Thomas W. Scott and Amy C. Morrison

Contents

Abstract Accounting for variation in mosquito vector populations will improve dengue surveillance and prevention. Because *Aedes aegypti*, the principle dengue virus (DENV) vector, transmit the virus with remarkable efficiency, entomological thresholds are especially low. Assessing risk of human infection based on immature mosquito indices has proven difficult. Greater emphasis should be placed on relative abundance of adult vectors in relation to human serotype-specific herd immunity, introduction of unique viruses, mosquito-human contact and weather. The most appropriate spatial scale for assessing entomological risk is the individual household. The scale for measuring DENV transmission risk has yet to be determined but is clearly larger than the household and likely to exceed several city blocks. Because households are expected to be a primary site for human DENV infection, intradomicile vector control strategies should be a priority, especially when the force of transmission is high. The most effective intervention strategy will

T.W. Scott (✉) and A.C. Morrison
Department of Entomology, University of California, Davis, CA, 95616, USA
e-mail: twscott@ucdavis.edu

A.L. Rothman (ed.), *Dengue Virus*, Current Topics in Microbiology and Immunology 338,   115
DOI 10.1007/978-3-642-02215-9_9, © Springer-Verlag Berlin Heidelberg 2010

combine vector control with vaccine delivery for rapid and sustained disease prevention.

1   Introduction

Effective vector-borne disease prevention requires understanding of dynamics in arthropod vector populations and application of that knowledge to interfere with or prevent pathogen transmission. A vector can impact transmission through differences in abundance, behavior, or genetics; all of which are factors that have well documented effects on the distribution of host infection and disease through time and space (Woolhouse et al. 1997; Lloyd-Smith et al. 2005). Theoretical and empirical studies of heterogeneities in vector-borne disease indicate that targeted monitoring and intervention have a greater impact on reducing disease than uniform applications across affected populations (Woolhouse et al. 1997; Lloyd-Smith et al. 2005; Smith et al. 2005). The public health implication of this observation is that uneven patterns of infection need to be accounted for in the design of disease surveillance and control programs. This is especially true in resource-limited endemic situations that often bear the greatest burden of disease. The challenge is to first define the relative contributions of different sources of heterogeneity and then determine if operationally relevant information on key components of variation can be obtained and sensibly applied. In this chapter we discuss how accounting for variation in vector populations can be applied to improve dengue prevention.

2   Dengue Virus Transmission

Transmission of all four dengue virus (DENV) serotypes is maintained by horizontal transfer in an *Aedes aegypti*-human cycle (Gubler 1989; Rodhain and Rosen 1997). Other mosquito species in the subgenus *Stegomyia* have been incriminated as vectors but *Ae. aegypti* is the most important DENV vector worldwide (Gubler and Kuno 1997). Mosquito infection begins when females imbibe viremic blood from a human host. After surviving an extrinsic incubation period of 7–14 days vectors become infectious and can transmit virus by bite (Watts et al. 1987). Within high and low extremes, at which vectors die or virus replication ceases, the duration of the extrinsic incubation is inversely related to ambient temperature. Humans incubate virus for 3–15 days (typically 4–7 days). Viremia can precede fever, lasts ~5 days and subsides with the loss of the ability to detect virus in blood (Focks et al. 1995; Waterman SH, Gubler D, 1989; Vaughn et al. 2000).

Unique features of its ecology and blood feeding behavior make *Ae. aegypti* a remarkably efficient DENV vector. Larvae and pupae develop in water held in artificial, often man-made, containers in and around human habitations (Gubler 1989). Adults tend to rest on clothing and surfaces inside homes where females frequently and almost exclusively bite human hosts (Scott et al. 1993b, 2000b).

Because human blood sources, mates and places to lay their eggs are abundant within the homes where they reside, adult *Ae. aegypti* tend not to disperse far (Morland and Hayes 1958; McDonald 1977; Trpis and Hausermann 1986; World Health Organization, 1999, 1997; Edman et al. 1998; Harrington et al. 2001a, b, 2005). In dengue-endemic settings few fly more than 100 m. For that reason, rapid geographic spread that can be characteristic of dengue outbreaks (Morrison et al. 1998) is most likely driven by movement of viremic humans, rather than flying infected mosquitoes.

Repeated feeding on human blood (i.e., every day or every other day) results in a fitness advantage for female *Ae. aegypti* and increases the efficiency with which they transmit virus (Scott et al. 1993a, 1997; Morrison et al. 1999; Harrington et al. 2001a). Blood feeding frequency is influenced both indirectly and directly by ambient temperatures. Smaller mosquitoes feed more often than larger ones (Scott et al. 2000b) and higher temperatures can augment immature development resulting in smaller mosquitoes. Higher temperatures also speed blood meal digestion so that females need to feed more often (Scott et al. 1997). Blood feeding frequency and thus contact rate between vectors and humans, appears to be facultative. It is influenced by the relative availability of human hosts and nectar sources (Edman et al. 1992; Van Handel et al. 1994; Martinez-Ibarra et al. 1997). In addition to temperature, body sizes of adult mosquitoes are determined by the resources available during larval development (Arrivillaga and Barrera 2004; Barrera et al. 2006). Because they live so close to human hosts and bite people so often, DENV transmission can occur even when *Ae. aegypti* population densities are low (Kuno 1995; Kuno 1997).

3 Dengue Prevention

Control of dengue is currently limited to decreasing *Ae. aegypti* population densities or preventing their contact with human hosts (Morrison et al. 2008). Even though encouraging progress has been made on the development of a DENV vaccine (Pediatric Dengue Vaccine Initiative funded by the Bill and Melinda Gates Foundation, http://www.pdvi.org) none are presently licensed and available for commercial use. Similarly, although development of anti-DENV drugs is underway (Farrar et al. 2007) their regular use in clinical settings does not appear imminent. Even after these important tools are in hand, vector control will continue to be a valuable weapon in the arsenal for combating dengue. Vector control has a broad public health impact and it complements other forms of disease prevention (Scott and Morrison 2008). Vector control reduces the risk of infection from a variety of viral pathogens other than DENV and reduces the proportion of people who need to be vaccinated by reducing the force of virus infection.

In many parts of the world, vector control strategies currently used for dengue are extensions of nearly century old programs designed to prevent yellow fever (Scott and Morrison 2008). Those approaches were based on the concept of vector eradication to terminate virus transmission. They demonstrated that rigorous adherence to elimination of vectors will significantly reduce or eliminate virus transmission, at least for the short term. They did not, however, provide the insight necessary to define quantitative relationships between mosquito abundance and DENV transmission, which is required if the goal is disease management through vector population suppression (PAHO 1994; Gubler and Kuno 1997; Reiter and Gubler 1997; Scott and Morrison 2003, 2008). Primarily because public health resources were limited and because of the difficulties of controlling dengue in growing urban endemic environments, in the early 1990s the Pan American Health Organization (PAHO) decided that the concept of *Ae. aegypti* eradication should be abandoned (Pan American Health Organization, 1994). The new public health policy for dengue prevention was "cost-effective utilization of limited resources to reduce vector populations to levels at which they are no longer of significant public health importance."

The new strategy is more complicated than the first and has been poorly defined in practice (Scott and Morrison 2008). The approach implies that *Ae. aegypti* population densities must be maintained close to or below minimum thresholds necessary for virus transmission. Reducing vector densities below thresholds slows the force of DENV transmission so that sequential infections with heterologous serotypes are similarly diminished and the incidence of serious disease will decrease or be eliminated (Vaughn et al. 2000). The latter assumes that reducing severe disease but not all disease, is a reasonable public health objective. The emphasis is on reducing vector population density, rather than decreasing vector life span or contact with human hosts; both of which are expected to have a greater impact on human infection than only reducing vector density (Morrison et al. 2008; Scott and Morrison 2008). The new approach requires (1) understanding of relationships between adult *Ae. aegypti* density, DENV transmission and disease incidence and (2) application of a surveillance system that can follow fluctuations in those processes and guide dynamic control programs in a timely fashion. This is a challenging proposition. Relationships between entomological measures of risk and human infection are not well enough understood for any arbovirus. The closest are studies on mosquito density and sentinel chicken seroconversion to western and St. Louis encephalitis viruses in southern California (Reeves 1971; Olson et al. 1979). Similar entomologic and human epidemiologic data are needed for dengue.

4 Mosquito Densities and Virus Transmission

4.1 Surveillance

Because vector eradication is no longer the goal, ongoing surveillance will be essential for the success of the new dengue public health policy. A more effective early warning system for dengue is in great demand (DeRoeck et al. 2003). It will be necessary to estimate, monitor and operationally respond to fluctuations in mosquito populations. Because targets for control (i.e., entomological thresholds) are dynamic and low from an operational perspective (Focks et al. 2000), their precise estimation has been a significant challenge. Thresholds can fluctuate depending on human serotype-specific herd immunity, introduction of different viruses, rate of mosquito-human contact and ambient temperature (Scott and Morrison 2003).

Due to the complex natural history of arboviruses like dengue, the most operationally feasible way to predict dynamic thresholds is with user friendly analytical models (i.e., Focks et al. 1993a,b, 1995; Hemingway et al. 2006; Morrison et al. 2008). The goal should be to use model results to prioritize and direct prevention to times and places where transmission risks are greatest. We expect that improved dengue surveillance and control will use this method to integrate more than entomological measures into surveillance and control.

Few attempts to predict dengue epidemics have been successful. When all 4 serotypes are not present, detection of a novel serotype introduction can be a powerful predictor of epidemic transmission (Gubler and Casta-Velez 1991). To date, sensitivity of a limited number of viral surveillance programs has been insufficient and novel serotypes are usually detected months after their introduction (Gubler 2002). In hyperendemic areas (i.e., all 4 serotypes are present) viral, serological or syndromic case surveillance can be more informative than monitoring vector densities alone. Even so, because of the residual eradication methodologies combined with financial restraints, uniformly prescribed entomological indices are often used to evaluate the potential for DV transmission and to assess effectiveness of vector control efforts (Focks and Chadee 1997; Focks et al. 2000; Scott and Morrison 2003; Scott and Morrison 2008).

4.2   Entomological Measures of Risk

A weak association between immature *Ae. aegypti* indices and DENV transmission has been repeatedly reported (Tun-Lin et al. 1995, 1996; Focks and Chadee 1997; Reiter and Gubler 1997; Scott and Morrison 2003; Kay and Nam 2005). Shortfalls are due to failure to account for larval mortality, heterogeneity in container productivity (Tun-Lin et al. 1995; Focks and Chadee 1997; Morrison et al. 2004b), temporal differences in *Ae. aegypti* life stages and variation in susceptibility of the human population to DENV infection (Scott and Morrison 2008). Assessments are further complicated by difficulties in dealing with heterogeneities in the distribution of immature *Ae. aegypti*, which operationally means that estimates of vector density are susceptible to sampling error (Tun-Lin et al. 1995, 1996; Focks and Chadee 1997; Getis et al. 2003; Morrison et al. 2004a, b).

The most commonly applied *Stegomyia* indices include the premise/house index (percentage of houses infested with larvae and/or pupae), container index (percentage of water-holding containers infested with larvae and/or pupae), Breteau index (number of positive containers per 100 houses) or oviposition trap (ovitrap) data, all of which were intended to detect the presence or absence of *Ae. aegypti*, not the relative abundance of adult virus transmitting mosquitoes (Conner and Monroe 1923; Breteau 1954; Tun-Lin et al. 1995; Focks and Chadee 1997). Alternatives to traditional entomological indices are larval (Chan et al. 1971; Bang et al. 1981; Tun-Lin et al. 1995, 1996) and pupal productivity surveys, both of which are intended to account for heterogeneity in production of adult mosquitoes from different container types (Focks and Chadee 1997). Traditional indices are less sensitive to sampling error than those that measure absolute numbers of larvae, pupae, or adult mosquitoes per household or unit area but are insensitive to epidemiologically relevant density fluctuations (Morrison et al. 2004). The most direct measure of transmission risk would be an index with 'person' in the denominator (e.g., pupae per person) (Dye 1992; Focks et al. 1995). Per person indices have been shown to have the highest coefficients of variation (Morrison et al. 2004) and estimation is complicated by obtaining accurate human population counts. Measurement of human populations is difficult because of their movement. The same individuals could be counted multiple times at the different places they visit during their daily activities and are at risk of DENV infection (Morrison et al. 2006). For example, school children are at risk at home and school. Working adults are at risk at home, their work place, markets, churches, etc. Consequently, we recommend that effective entomological surveillance not rely on any single indicator, rather it should be based on careful interpretation of a combination of indices.

Another component of entomological surveillance is productivity analysis; that is, local characterization of important larval habits. One method, the pupal demographic survey, determines the relative importance of different container types by quantifying pupal production by container category (Focks et al. 1993a,b; Focks and Chadee 1997). Application of this technique is currently being evaluated for targeted vector control (Focks and Alexander 2006; World Health Organization, 2006), that is, directing control to immature *Ae. aegypti* development sites that produce most of the adults, rather than uniform or indiscriminate treatment of sites. Issues of sampling error discussed above are applicable to these methodologies; i.e., large samples are best. Rigorous field studies are needed to test the hypothesis that removal of key production sites results in predicted population reductions and that accounting for adult production will improve dengue prevention (Morrison et al. 2004).

What is sorely needed are means of assessing adult *Ae. aegypti* density (Morrison et al. 2008; Scott and Morrison 2008). Because they do not readily enter commonly used mosquito traps, adult *Ae. aegypti* are difficult to collect in a surveillance context (Jones et al. 2003). Moreover, their population densities tend to be low (Scott et al. 2000a), which further complicates estimating population sizes. Adult population density as determined by mosquitoes collected with backpack aspirators have, however, been correlated with the incidence of human DENV infection (Morrison and Scott, unpublished data). Immature indices can be useful for characterizing spatial patterns of *Ae. aegypti* infestations and have been positively associated with human dengue seroprevalence but only adult density has been linked to the epidemiologically important incident rate. Adult trap development should be a top priority for dengue surveillance improvement.

4.3 Spatial Scale for Surveillance and Control

Because results of risk prediction can vary across different geographic scales, it is critically important to determine the proper scale at which to carry out *Ae. aegypti* surveillance (Scott and Morrison 2008). Ecologists have traditionally focused on characterizing temporal variation in *Ae. aegypti* density (Sheppard et al. 1969; Gould et al. 1970; Yasuno and Pant 1970). At our study site in Iquitos, Peru, we determined that entomological risk for dengue is best measured frequently at a small-scale – at the level of a household (Getis et al. 2003; Morrison et al. 2004a). Households represent a statistically independent unit; infested households are randomly distributed within communities and their locations change over time (Getis et al. 2003). Adult *Ae. aegypti* cluster most strongly within individual houses but clusters can extend to 30 meters. Because infested houses are not normally distributed, sample sizes for *Ae. aegypti* need to be large in order to adequately assess risk and evaluate intervention effectiveness (Alexander et al. 2006).

It is worth noting that even though households, or residential sites, should be the focus of surveillance and control, non-residential sites (e.g., schools, factories, ports, public markets, parks, commercial zones) merit attention, too. *Ae. aegypti* surveillance and control programs should not treat them as a residential site (Morrison et al. 2006). The relative importance of these sites should be assessed locally, so that customized situation-specific surveillance and control strategies can be developed. Even when infestations are low, some of these sites can be extremely productive with important impacts on neighboring communities. For sites where people gather (e.g., schools or cemeteries) even low densities of mosquitoes can expose many people to DENV.

Spatial analyzes of entomological and epidemiological variables will be necessary for addressing the challenges of dengue surveillance because they will reveal patterns of DENV transmission that can be used to design progressively more effective intervention strategies (Eisen and Lozano-Fuentes 2009). We can ask, for example, whether clusters of human dengue cases are due primarily to variation in *Ae. aegypti* population density or some other factor(s) inherent to a particular study area (Gatrell et al. 1996). Foci of DENV transmission can be identified and associated with certain locations in the community, like neighbor's homes, local meeting or gathering places. We know that DENV infections can cluster in households (Morrison et al. 1998) and speculate that it is due to *Ae. aegypti*'s restricted flight range and propensity to bite people frequently; a single infective mosquito can transmit virus to numerous people in a short period of time. What we do not know is how daily human movements affect the patterns of DENV infection or the details of DENV spread from households across larger geographic scales. Consequently, one of the more important contributions of spatial analyzes to dengue surveillance and control will be to define the proper geographic scale for predicting entomological risk and the threat of DENV transmission. Based on our analyzes to date, we expect that virus transmission needs to be measured at larger scales than for entomological risk (Scott and Morrison 2003, 2008).

4.4   Intradomicile Vector Control

Focusing intervention on the primary point of human exposure to virus should be an emphasis of dengue prevention. Because *Ae. aegypti* rest, blood feed, mate and reproduce in houses (Scott et al. 2000b), homes are expected to be a principal site for human exposure to DENV (DeBenedictis et al. 2003; Garcia-Rejon et al. 2008) and, thus, a focus of efforts to break transmission (Morrison et al. 2008). It is well established that if insecticides are not delivered inside houses they are ineffective because outdoor applications do not reach adult *Ae. aegypti* where they rest (Reiter and Gubler 1997). Trial results from intradomicile application of indoor residual sprays (IRS) and insecticide treated materials (ITM) are encouraging because they reduced *Ae. aegypti* populations and human infections (Nam et al. 1993; Nguyen et al. 1996; Igarashi 1997; Kroeger et al. 2006). Moreover, intradomicile adulticide interventions do more than decrease adult mosquito density. They reduce vector lifespan, which has a greater overall impact on virus transmission than reducing vector density alone. In addition, intradomicile control will decrease other vector-borne diseases and insect pest problems (Morrison et al. 2008). Even if adult mosquito populations recover quickly, indoor application of insecticides can interrupt epidemic virus transmission, resulting in a noticeable impact on disease (Morrison and Scott, unpublished data). Residual insecticides are a way to protract the adult killing effect. Consequently, intradomicile control can be leveraged into cost- and operationally-effective public health programs with broad impacts. A key component of this approach will be development of improved means for detecting and managing insecticide resistance in *Ae. aegypti* and other targeted insect populations (Hemingway et al. 2006).

4.5   Paradigm Shift to Local Level Surveillance and Prevention

Improving dengue prevention requires a conceptual shift away from current methods that emphasize broadly prescribed and applied strategies to local public health officials and control personnel deciding themselves the best strategy for their particular situation. We expect that adaptive disease management will be the basis for improving contemporary dengue prevention (Scott and Morrison 2008). This will require the capacity to account for variation through time and space in mosquito vector dynamics and virus transmission. Local personnel will need to regularly evaluate and adapt their surveillance and response programs. A conceptual framework will need to be developed that can be applied across different ecological and epidemiologic situations to determine (1) what control procedures should be used; (2) how they should be applied; and (3) how they should be evaluated and/or monitored. To do that will require answers to questions like: What should the site and situation-specific goal(s) be for dengue prevention programs? How should control be monitored (i.e., what surveillance and risk assessment programs should be used)? What disease prevention tools are effective and currently available and which ones needed to be developed? What are the best integrated and adaptive control programs (e.g., dynamic application of vector control in concert with other disease prevention and management strategies)? What major steps need to be taken to develop, evaluate, disseminate and ensure application of effective and sustainable dengue prevention? Development of an adaptive dengue management system is one of the greatest public health needs and biggest challenges for contemporary dengue researchers.

4.6 Vector Control and Vaccines

Although entomologists have recognized for some time that integrated, multidimensional control strategies are superior to a single line of attack (Shea et al. 2000), combinations of tactics have largely been drawn from within disciplines (WHO 2006). For example, using combinations of different vector control methods such as source reduction, larvacides and adulticides. We advocate an expansion of this idea to include the integration of vector control and vaccines (Scott and Morrison 2008). Synergistic benefits of vector control and chemotherapy have been demonstrated for control of lymphatic filariasis (Sunish et al. 2007). The overall goal in the combined strategy is to reduce the basic reproductive rate (R_o) of dengue below 1. A secondary aim is to reduce severe disease by decreasing infection incidence. The two approaches complement one another. Vector control reduces the force of pathogen transmission, which lowers the critical proportion of the human population that needs to be vaccinated (Anderson and May 1991). With an integrated strategy, reduced vaccine delivery goals are more operationally realistic. On the other hand, successful vector control is difficult to sustain because as mosquito densities decline and virus transmission is reduced over time the proportion of susceptible people increases (i.e., recruitment of susceptibles by birth), which in turn drives down entomological thresholds. Meeting lower thresholds becomes incrementally more difficult to achieve. A vaccine, however, can be used to ease the task by elevating herd immunity so that entomological thresholds remain relatively high. In combination with a vaccine, sustained, effective vector control becomes an operationally achievable task. When applied together vector control and vaccines have the potential for more swift and prolonged dengue prevention than if either approach is used by itself.

5 Conclusions

Improved dengue prevention will require (1) a better understanding of variation in the relationship between vectors and infected humans and (2) a conceptual shift from universal to locally derived strategies for surveillance and control. Measures of entomological risk and vector control tactics should change from an emphasis on immature *Ae. aegypti* (eggs, larvae and pupae) to adult female mosquitoes, the life stage that transmits DENV. Adaptive disease management will require that public health officials and vector control personnel are provided the necessary tools to interpret local surveillance data, design the most appropriate situation-specific intervention plans and monitor the impact of their mosquito and disease management efforts. Although nonresidential sites may under certain circumstances be important for overall patterns of transmission, emphasis on intradomicile interventions are most likely to reduce virus transmission by attacking dengue at the primary site of human infection. An improved understanding of spatial heterogeneities in human infection will support strategic and cost-effective surveillance and disease prevention. Integration of intervention methods, especially those that cut across disciplines by combining vaccines and vector control, will dramatically improve the effectiveness and more importantly the sustainability of dengue prevention programs.

References

Alexander N, Lenhart AE, Romero-Vivas CME, Barbazan P, Morrison AC, Barrera R, Arredondo-Jime Nez JI, Focks DA (2006) Sample sizes for identifying the key types of container occupied by dengue-vector pupae: the use of entropy in analyses of compositional data. Ann Trop Med Parasit 100:S5–S16

Anderson R, May R (1991) Infectious Diseases of Humans. Oxford University Press, Oxford, UK, p 735

Arrivillaga J, Barrera R (2004) Food as a limiting factor for *Aedes aegypti* in water-storage containers. J Vector Ecology 29:11–20

Bang YH, Brown DN, Onwubiko AO (1981) Prevalence of potential yellow fever vectors in domestic water containers in south-east Nigeria. Bull WHO 59:107–114

Barrera R, Amador M, Clark GG (2006) Use of the pupal survey technique for measuring *Aedes aegypti* (Diptera : Culicidae) productivity in Puerto Rico. Am J Trop Med Hyg 74:290–302

Breteau H (1954) La fievere jaune en Afrique occidentale francaise. Un aspect de la medicine preventive massive. Bull WHO 11:453–481

Chan YC, Chan KL, Ho BC (1971) *Aedes aegypti* (L.) and *Aedes albopictus* (Skuse) in Singapore: I. Distribution and density. Bull WHO 44:617–627

Conner ME, Monroe WM (1923) *Stegomyia* indices and their value in yellow fever control. Am J Trop Med 4:9–19

DeBenedictis J, Chow-Schaffer E, Costero A, Clark GG, Edman JD, Scott TW (2003) Identification of the people from whom engorged *Aedes aegypti* took blood meals in Florida, Puerto Rico using PCR-based DNA profiling. Am J Trop Med Hyg 68:447–452

DeRoeck D, Jacqueline D, Clemens JD (2003) Policymakers' views on dengue fever/dengue haemorrhagic fever and the need for dengue vaccines in four southeast Asian countries. Vaccine 22:121–129

Dye C (1992) The analysis of parasite transmission by bloodsucking insects. Ann Rev Entomol 37:1–19

Edman JD, Scott TW, Costero A, Morrison AC, Harrington LC, Clark GG (1998) *Aedes aegypti* (L.) (Diptera: Culicidae) movement influenced by availability of oviposition sites. J Med Entomol (Traub Memorial) 35:578–583

Eisen L, Lozano-Fuentes S (2009) Use of mapping and spatial and space-time modeling approaches in operational control of Aedes aegypti and dengue. PLoS Negl Trop Dis 3:e411

Farrar J, Focks, D, Gubler, D, Barrera, R, Guzman, MG, Simmons, C, Kalayanarooj, S, Lum, L, McCall, PJ, Lloyd, L, Horstick, O, Dayal-Drager, R, Nathan, MB, Kroeger A (2007) Towards a global dengue research agenda. Trop Med Internat Hlth 12:695–699 On behalf of the WHO/TDR Dengue scientific working group

Focks DA, Haile DG, Daniels E, Mount GA (1993a) Dynamic life table model for *Aedes aegypti* (L.) (Diptera: Culicidae). Analysis of the literature and model development. J Med Entomol 30:1003–1017

Focks DA, Haile DG, Daniels E, Mount GA (1993b) Dynamic life table model for *Aedes aegypti* (L.) (Diptera: Culicidae). Simulation results and validation. J Med Entomol 30:1018–1028

Focks DA, Daniels E, Haile DG, Keesling JE (1995) A simulation model of the epidemiology of urban dengue fever: literature analysis, model development, preliminary validation, and samples of simulation results. Am J Trop Med Hyg 53:489–506

Focks DA, Chadee DD (1997) Pupal survey: an epidemiologically significant surveillance methods for *Aedes aegypti*. An example using data from Trinidad. Am J Trop Med Hyg 56:159–167

Focks DA, Brenner RJ, Hayes J, Daniels E (2000) Transmission thresholds for dengue in terms of Aedes aegypti pupae per person with discussion of their utility in source reduction efforts. Am J Trop Med Hyg 62:11–18

Focks DA, Alexander N (2006) M*ulticountry study of Aedes aegypti pupal productivity survey methodology: findings and recommendations.* World Health Organization, Geneva, Switzerland

Garcia-Rejon J, Loroño-Pino MA, Farfan-Ale JA, Flores-Flores L, Del Pilar Rosado-Paredes E, Rivero-Cardenas N, Najera-Vazquez R, Gomez-Carro S, Lira-Zumbardo V, Gonzalez-Martinez P, Lozano-Fuentes S, Elizondo-Quiroga D, Beaty BJ, Eisen L (2008) Dengue virus-infected Aedes aegypti in the home environment. Am J Trop Med Hyg 79:940–950

Gatrell AC, Bailey TC, Diggle PJ, Rowlingson BS (1996) Spatial point pattern analysis and its application in geographical eoidemiology. Trans Institute British Geographers 21:256–274

Getis A, Morrison AC, Gray K, Scott TW (2003) Characteristics of the Spatial Pattern of the dengue vector, *Aedes aegypti*, in Iquitos, Peru. Am J Trop Med Hyg 69:494–503

Gould DJ, Mount GA, Scanlon JE, Ford HR, Sullivan MF (1970) Ecology and control of dengue vectors on an island in the Gulf of Thailand. J Med Entomol 7:499–508

Gubler DJ (1989) Dengue. In: Monath TP (ed) The Arboviruses: epidemiology and ecology, vol 2. CRC Press, Boca Raton, FL, pp 223–260

Gubler DJ, Casta-Velez A (1991) A program for prevention and control of epidemic dengue and dengue hemorrhagic fever in Puerto Rico and the U.S. Virgin Islands. Bull PAHO 25:237–247

Gubler DJ, Kuno G (1997) Dengue and Dengue Hemorrhagic fever. CAB International, New York, p 462

Gubler DG (2002) How effectively is epidemiological surveillance used for dengue programme planning and epidemic response? Dengue Bull 26:96–106

Harrington LC, Edman JD, Scott TW (2001a) Why do female *Aedes aegypti* (Diptera: Culicidae) feed preferentially and frequently on human blood? J Med Entomol 38:411–422

Harrington LC, Buonaccorsi JP, Edman JD, Costero A, Clark GG, Kittayapong P, Scott TW
    (2001b) Analysis of survival rates for two age cohorts of *Aedes aegypti* (L.) (Diptera:
    Culicidae): Results from Puerto Rico and Thailand. J Med Entomol 38:537–547
Harrington LC, Scott TW, Lerdthusnee K, Coleman RC, Costero A, Clark GG, Jones JJ, Kitthawee
    S, Kittayapong P, Sithiprasasna R, Edman JD (2005) Dispersal of the dengue vector *Aedes
    aegypti* within and between rural communities. Am J Trop Med Hyg 72:209–220
Hemingway J, Beaty BJ, Rowland M, Scott TW, Sharp BL (2006) The Innovative Vector Control
    Consortium: Improved control of mosquito-borne diseases in and around the home. Trends in
    Parasitology 22:308–312
Igarashi A (1997) Impact of dengue virus infection and its control. FEMS Immunol and Med
    Mcrobiol 18:291–300
Martinez-Ibarra JA, Rodriguez MH, Arredondo-Jimenez JI, Yuval B (1997) Influence of plant
    abundance on nectar feeding by *Aedes aegypti* (Diptera: Culicidae) in southern Mexico. J Med
    Entomol 34:589–593
Jones JW, Sithiprasasna R, Schleich S, Coleman RE (2003) Evaluation of selected traps as tools
    for conducting surveillance for adult Aedes aegypti in Thailand. J Amer Mosq Control Assoc
    19:148–150
Kay B, Nam VS (2005) New strategy against *Aedes aegypti* in Vietnam. Lancet 365:613–617
Kroeger A, Lenhart A, Ochoa M, Villegas E, Levy M, Alexander N, McCall PJ (2006) Effective
    control of dengue vectors with curtains and water container covers treated with insecticide in
    Mexico and Venezuela: cluster randomised trials. BMJ 332:1247–1252
Kuno G (1995) Review of the factors modulating dengue transmission. Epidemiol Reviews
    17:321–335
Kuno G (1997) Factors influencing the transmission of dengue viruses. In: Gubler DJ, Kuno G
    (eds) Dengue and Dengue Hemorrhagic Fever. CAB International, New York, pp 61–88
Lloyd-Smith JO, Schreibe SJ, Kopp PE, Getz WM (2005) Superspreading and the effect of
    individual variation on disease emergence. Nature 438:355–359
Mcdonald PT (1977) Population characteristics of domestic *Aedes aegypti* (Diptera: Culicidae) in
    villages on the Kenya Coast II. Dispersal within and between villages. J Med Entomol 14:49–53
Morland HB, Hayes RO (1958) Urban dispersal and activity of *Aedes aegypti*. Mosq News
    18:137–144
Morrison AC, Getis A, Santiago M, Rigua-Perez JG, Reiter P (1998) Exploratory space-time
    analysis of reported dengue cases during an outbreak in Florida, Puerto Rico, 1991–1992.
    Am J Trop Med Hyg 58:287–298
Morrison AC, Costero A, Edman JD, Scott TW (1999) Increased fecundity of female *Aedes
    aegypti* (L.) (Diptera: Culicidae) fed only human blood prior to release in Puerto Rico.
    J Am Mosq Control Assoc (Barr issue) 15:98–104
Morrison AC, Gray K, Getis A, Estete H, Sihuincha M, Focks D, Watts D, Scott TW (2004a)
    Temporal and geographic patterns of *Aedes aegypti* (Diptera: Culicidae) production in Iquitos,
    Peru. J Med Entomol 41:1123–1142
Morrison AC, Astete H, Chapilliquen F, Dias G, Gray K, Getis A, Scott TW (2004b) Evaluation of
    a sampling methodology for rapid assessment of *Aedes aegypti* infestation levels in Iquitos.
    J Med Entomol 41:502–510
Morrison AC, Sihuincha JD M, Stancil E, Zamora H, Astete JG Olson, Vidal-Ore C, Scott TW
    (2006) *Aedes aegypti* (Diptera:Culicidae) production from non-residential sites in the Amazo-
    nian city, Iquitos, Peru. Ann Trop Med Parasit 100:S73–S86
Morrison AC, Zielinski-Gutierrez E, Scott TW, Rosenberg R (2008) Defining the challenges and
    proposing new solutions for *Aedes aegypti*-borne disease prevention. PLoS Medicine 5:362–366
Nam VS, Nguyen HT, Tien TV, Niem TS, Hoa NT, Thao NT, Tron TQ, Yen NT, Ninh TU, Self LS
    (1993) Permethrin-treated bamboo curtains for dengue vector control-field trial, Viet Nam.
    Dengue Newsletter 18:23–28
Nguyen HT, Tien TV, Tien HC, Ninh TU, Hoa NT (1996) The effect of Olyset net screen to
    control the vector of dengue fever in Vietnam. Dengue Bulletin 20:87–92

Olson JG, Reeves WC, Emmons RW, Milby MM (1979) Correlation of *Culex tarsalis* population indices with the incidence of St. Louis encephalitis and western equine encephalomyclitis in California. Am J Trop Med Hyg 28:335–343

Pan American Health Organization (1994) Dengue and Dengue Hemorrhagic Fever in the Americas. Guidelines for Prevention and Control. Pan American Health Organization, Washington DC Pan American Health Organization Scientific Publication no. 548

Reeves WC (1971) Mosquito vector and vertebrate host interaction: the key to maintenance of certain arboviruses. In: Fallis AM (ed) Ecology and physiology of parasites. Toronto, ON, University of Toronto Press, pp 223–230

Reiter P, Gubler DJ (1997) Surveillance and control of urban dengue vectors. In: Gubler DJ, Kuno G (eds) Dengue and Dengue Hemorrhagic fever. CAB International, New York, pp 425–462

Rodhain F, Rosen L (1997) Mosquito vectors and dengue virus-vector relationships. In: Gubler DJ, Kuno G (eds) Dengue and Dengue Hemorrhagic fever. CAB International, New York, NY, pp 61–88

Scott TW, Clark GG, Lorenz LH, Amerasinghe PH, Reiter P, Edman JD (1993a) Detection of multiple blood-feeding by *Aedes aegypti* during a single gonotrophic cycle using a histological technique. J Med Entomol 30:94–99

Scott TW, Chow E, Strickman D, Kittayapong P, Wirtz RA, Edman JD (1993b) Bloodfeeding patterns of *Aedes aegypti* in a rural Thai village. J Med Entomol 30:922–927

Scott TW, Naksathit A, Day JF, Kittayapong P, Edman JD (1997) Fitness advantage for *Aedes aegypti* and the viruses it transmits when females feed only on human blood. Am J Trop Med Hyg 52:235–239

Scott TW, Morrison AC, Lorenz LH, Clark GG, Strickman D, Kittayapong P, Zhou H, Edman JD (2000a) Longitudinal studies of *Aedes aegypti* (L.) (Diperta: Culicidae) in Thailand and Puerto Rico: Population dynamics. J Med Entomol 37:77–88

Scott TW, Amerasinghe PH, Morrison AC, Lorenz LH, Clark GG, Strickman D, Kittayapong P, Edman JD (2000b) Longitudinal studies of *Aedes aegypti* (L.) (Diperta: Culicidae) in Thailand and Puerto Rico: Blood feeding frequency. J Med Entomol 37:89–101

Scott TW, Morrison AC (2003) *Aedes aegypti* density and the risk of dengue virus transmission. In: Takken W, Scott TW (eds) Ecological aspects for application of genetically modified mosquitoes. FRONTIS, Dordrecht, The Netherlands, pp 187–206

Scott TW, Morrison AC (2008) Longitudinal field studies will guide a paradigm shift in dengue prevention in. Vector-borne diseases: understanding the environmental, human health, and ecological connections. Washington DC, The National Academies Press, pp 132–149

Shea P, Thrall H, Burdon JJ (2000) An integrated approach to management in epidemiology and pest control. Ecol Lett 3:150–158

Sheppard PM, Macdonald WW, Tonn RJ, Grab B (1969) The dynamics of an adult population of *Aedes aegypti* in relation to dengue haemorrhagic fever in Bangkok. J Anim Ecol 38:661–702

Smith DL, Dushoff J, Snow RW, Hay SI (2005) The entomological inoculation rate and *Plasmodium falciparum* infection in African children. Nature 438:492–495

Sunish IP, Rajendran R, Mani TR, Munirathinam A, Dash AP, Tyagi BK (2007) Vector control complements mass drug administration against bancroftian filariasis in Tirukoilur, India. Bull WHO 85:138–145

Trpis M, Hausermann W (1986) Dispersal and other population parameters of *Aedes aegypti* in an African village and their possible significance in epidemiology of vector-borne diseases. Am J Trop Med Hyg 35:1263–1279

Tun-Lin W, Kay BH BH, Barnes A (1995) Understanding productivity, a key to *Aedes aegypti* surveillance. Am J Trop Med Hyg 53:595–601

Tun-Lin W, Kay BH, Barnes A, Forsyth S (1996) Critical examination of *Aedes aegypti* indices: correlations with abundance. Am J Trop Med Hyg 53:595–601

Van Handel E, Edman JD, Day JF, Scott TW, Clark GG, Reiter P, Lynn HC (1994) Plant sugar, glycogen, and lipid assay of *Aedes aegypti* collected in urban Puerto Rico and rural Florida. J Amer Mosq Control Assoc 10:149–153

Vaughn DW, Green S, Kalayanarooj S, Innis BL, Nimmannitya S, Suntayakorn S, Endy TP, Raengsakulrach B, Rothman AL, Ennis FA, Nisalak A (2000) Dengue viremia titer, antibody response pattern, and virus serotype correlate with disease severity. J Infec Dis 181:2–9

Waterman SH, Gubler D (1989) Dengue Fever. Clin Dermatol 7:117–122

Watts DM, Burke DS, Harrison BA, Whitemire R, Nisalak A (1987) Effect of temperature on the vector efficiency of *Aedes aegypti* for dengue 2 virus. Am J Trop Med Hyg 36:143–152

Woolhouse MEJ, Dye C, Etard J-F, Smith T, Charlwood JD, Garnett GP, Hagan P, Hii JLK, Ndhlovu PD, Quinnell RJ, Watts CH, Chandiwana SK, Anderson RM (1997) Heterogeneities in the transmission of infectious agents: Implications for the design of control programs. PNAS 94:338–342

World Health Organization (1997) Dengue haemorrhagic fever: diagnosis, treatment, prevention and control. World Health Organization, Geneva, p 84

World Health Organization (1999) Prevention and Control of Dengue and Dengue Haemorrhagic Fever: Comprehensive Guidelines. WHO Regional Publication, SEARO No. 29. pp 134

World Health Organization (2006) Multicountry study of Aedes aegypti pupal productivity survey methodology: findings and recommendations. Switzerland, Geneva

Yasuno M, Pant C (1970) Seasonal changes in biting and larval infestation rates of *Aedes aegypti* in Bangkok, Thailand in 1969. Bull WHO 43:319–325

Dengue Vaccine Candidates in Development

Anna P. Durbin and Stephen S. Whitehead

Contents

Abstract Each of the DENV serotypes can cause the full spectrum of dengue illness. Epidemiological studies have implicated preexisting heterotypic DENV antibody as a risk factor for more severe disease upon secondary DENV infection. For these reasons, a successful DENV vaccine must protect against all four DENV serotypes. Live attenuated DENV vaccine candidates are the furthest along in development and clinical evaluation. Two live attenuated tetravalent vaccine candidates are in Phase 2 clinical trials in DENV endemic regions. Numerous other vaccine candidates including inactivated whole virus, recombinant subunit protein, DNA and virus-vectored vaccines are also under development. Those DENV

A.P. Dublin (✉)
Department of International Health, Johns Hopkins Bloomberg School of Public Health, USA
e-mail: adurbin@jhsph.edu

S.S. Whitehead
Laboratory of Infectious Diseases, National Institute of Allergy and Infectious Diseases, National Institutes of Health, USA

A.L. Rothman (ed.), *Dengue Virus*, Current Topics in Microbiology and Immunology 338,   129
DOI 10.1007/978-3-642-02215-9_10, © Springer-Verlag Berlin Heidelberg 2010

vaccine candidates that have been evaluated in preclinical animal models or in clinical trials will be discussed.

1 Introduction

Unlike other flaviviruses such as yellow fever virus, Japanese encephalitis virus and tick-borne encephalitis virus, no licensed vaccine for dengue exists. The many hurdles to the development of a successful DENV vaccine include the lack of an animal model that reproduces human disease, the need to develop a separate vaccine for each DENV serotype and the risk of inducing enhanced disease upon subsequent natural infection if antibody to one or more serotypes wanes over time. Therefore, an effective DENV vaccine must induce long-lasting and protective immunity against all four DENV serotypes (Whitehead et al. 2007). Despite these many obstacles, DENV vaccine development has made great strides in recent decades. There are two tetravalent vaccine candidates currently in Phase 2 clinical trials, several live vaccine candidates in Phase I clinical trials and many subunit, DNA and vectored vaccines in preclinical stages of development. The candidate vaccines that have been most extensively evaluated and are furthest along in development are live attenuated DENV vaccines.

2 Biologically-Derived Live Attenuated Dengue Vaccines

2.1 Mouse-Brain Derived Dengue Viruses

The earliest studies of live attenuated DENV vaccines were made in the 1940s by both American and Japanese investigators (Hotta 1952; Sabin 1952; Sabin and Schlesinger 1945; Schlesinger et al. 1956). These investigators serially passaged DENV-1 or DENV-2 by intra-cerebral inoculation of mice. As the virus became adapted to mice, it became less pathogenic for humans. Sabin demonstrated that the virus was attenuated for humans after the seventh mouse brain passage (Sabin 1952). Most of the volunteers vaccinated with the DENV mouse-brain suspension developed only mild symptoms of dengue, usually accompanied by a macular rash. In general, these vaccines induced neutralizing antibody against the vaccine strain (Wisseman et al. 1963) and protected volunteers from challenge with wild-type virus (Hotta 1952; Sabin 1952; Sabin and Schlesinger 1945; Schlesinger et al. 1956). Interestingly, Sabin and others noted decreased protection and/or lower neutralizing antibody titers induced by the DENV-1 vaccine when it was mixed with the live attenuated yellow fever 17D vaccine (Fujita et al. 1969; Sabin 1952). Hotta attempted to further attenuate the mouse adapted DENV-1 virus by mixing it with 5% ox-bile, which incompletely inactivated the virus (Hotta 1969). Twelve volunteers who received the vaccine were thought to be protected as they remained healthy during the large dengue outbreak in Osaka, Japan, in 1945. Due to concerns of impurities in the mouse-brain derived and ox-bile treated vaccines, these vaccines were not pursued further.

2.2 Tissue-Culture Derived Live Attenuated Dengue Viruses

The Walter Reed Army Institute of Research (WRAIR) in the United States and the Center for Vaccine Development at Mahidol University in Thailand began live attenuated dengue vaccine programs in the early 1980s. Candidates from both groups were developed from viruses isolated from dengue patients and then attenuated by sequential passage in primary dog kidney (PDK) cells or primary green monkey kidney (PGMK) cells. The Mahidol live attenuated DENV vaccine candidates identified for inclusion in a tetravalent formulation were DENV-1 (16007, PDK 13); DENV-2 (16681, PDK-53), DENV-3 (16562, PGMK-30, FRhL-3); and DENV-4 (1036 PDK-48). These vaccine candidates were tested as monovalent, bivalent and trivalent formulations in adult volunteers (Bhamarapravati and Yoksan 1989, 1997; Bhamarapravati et al. 1987; Vaughn et al. 1996). In general, these vaccines were well tolerated with fever, rash and mild liver enzyme elevations being the most commonly reported side effects. Seroconversion rates were reported as 90–100% for these candidates when tested as monovalent, bivalent, or trivalent formulations (Bhamarapravati and Sutee 2000; Bhamarapravati and Yoksan 1989). Importantly, those volunteers with preexisting dengue antibody who were vaccinated with a monovalent vaccine did not show increased severity of vaccine reactions. In addition, monotypic antibody remained detectable for up to 3 years, after which heterotypic antibody was detected, suggesting these vaccinees may have experienced a subsequent natural DENV infection (Bhamarapravati and Yoksan 1997). The tetravalent vaccine was then licensed to Aventis Pasteur (now Sanofi Pasteur) for further evaluation in a tetravalent formulation. Reactogencity appeared to be greater when the individual candidate vaccines were combined in a tetravalent formulation than when tested as monovalent vaccines (Kanesa-thasan et al. 2001). In addition, there appeared to preferential replication of the DENV-3 component, which also induced the highest neutralizing antibody titer. Due to possible interference of the DENV-3 component with the other serotypes, different dose formulations of the vaccine were further studied in adult volunteers. In addition, a two-dose vaccination schedule was evaluated for the first time (Sabchareon et al. 2002). Varying the concentration of DENV-3 in the tetravalent formulation and giving a second vaccination 6 months after the first improved the safety profile of the tetravalent vaccine and overcame the immunodominance of the DENV-3 component. The vaccine was then studied in 82 Thai children ages 5–12 years (Sabchareon et al. 2004). The vaccine was administered at three time points; day 0, 3–5 months later and again at 12 months after initial vaccination. The first dose of vaccine did not appear to enhance the severity of reactions with the second and third doses, despite low seroconversion rates to all four serotypes after the first vaccination. The most frequently reported adverse events were fever, rash and headache. Mild increases in liver enzymes and neutropenia were also reported. The frequency and severity of systemic reactions were reduced with subsequent doses of vaccine. Further trials of the Mahidol vaccine candidate were halted due to reactogenicity and formulation issues with the DENV-3 component (Kitchener et al. 2006).

WRAIR has generated numerous different DENV1-4 live vaccine candidates derived from clinical isolates by sequential passage in PDK cells. Several of these, although attenuated in preclinical studies, were subsequently found to be unacceptably reactogenic in human trials (Bancroft et al. 1984, 1981; Eckels et al. 1984; Innis et al. 1988; McKee et al. 1987; Scott et al. 1983). The DENV1-4 selected for initial inclusion in the WRAIR live attenuated DENV vaccine formulation were DENV-1 45AZ5 PDK-20, DENV-2 S16803 PDK-50, DENV-3 CH53489 PDK-20 and DENV-4 341750 PDK-20. These candidate viruses were tested first as monovalent formulations and then further evaluated in several different tetravalent formulations to determine the ideal passage number and dose of each candidate required to elicit the optimum safety and immunogenicity profile (Edelman et al. 1994; 2003; Kanesa-Thasan et al. 2003; Sun et al. 2003). The most common side effects noted in recipients of the tetravalent formulations tested were low-grade fever, rash and transient neutropenia (Edelman et al. 2003; Sun et al. 2003). A phase I trial evaluating this tetravalent vaccine in flavivirus-naïve children was recently completed (Simasathien et al. 2008). The tetravalent formulation comprised of DENV-1 45AZ5 PDK-27 at a dose of 6.1 $\log_{10}$ pfu, DENV-2 S16803 PDK-50 at a dose of 6.2 $\log_{10}$ pfu, DENV-3 CH53489 PDK-20 at a dose of 5.1 $\log_{10}$ pfu and DENV-4 341750 PDK-6 at a dose of 6.3 $\log_{10}$ pfu was tested in seven children ages 6–9 years in Bangkok. Two doses of the vaccine were administered 6 months apart. The vaccine was poorly immunogenic after dose 1, with the poorest antibody responses to DENV-1 but a tetravalent neutralizing antibody response was induced in all volunteers following the second dose of vaccine (Simasathien et al. 2008). The vaccine has been licensed by GlaxoSmithKline and is currently in Phase 2 clinical trials in Thailand.

2.3 Recombinant Live Virus Vaccines Attenuated by Engineered Mutations

DENV can be attenuated by the accumulation or introduction of mutations into the genome. Passage of DENV in PDK cells has led to the accumulation of mutations associated with an attenuation phenotype and specific mutations derived by this empirical approach were identified in 2000 as contributing to the attenuation of the DENV-2 PDK-53 vaccine candidate (Butrapet et al. 2000). In a separate vaccine strategy developed at the Laboratory of Infectious Diseases (NIAID, NIH, Bethesda, Maryland), the DENV-4 full-length cDNA clone was used to engineer deletion mutations into the 3'-UTR of DENV-4, which conferred varying levels of attenuation in rhesus monkeys compared to the wild-type parent virus (Men et al. 1996). Of particular interest was the 3' 172-143 deletion mutation, later referred to as Δ30, which maintained a desirable balance between level of attenuation and immunogenicity in monkeys. The DENV-4 virus containing the Δ30 mutation (DEN4Δ30) was subsequently evaluated in adult human volunteers and was shown to be safe, asymptomatic and immunogenic at all doses administered (10^1-10^5 pfu) (Durbin et al. 2001, 2005).

The success of the DEN4Δ30 vaccine in humans supported a unique strategy to create vaccine candidates for the other three DENV serotypes. Since the structure of the DENV- 3'-UTR is well conserved among all four serotypes, it was reasoned that deletion of nucleotides analogous to the Δ30 mutation in each serotype would likely result in attenuation. Introduction of the Δ30 mutation into DENV-1 resulted in a vaccine candidate attenuated to levels similar to that observed in monkeys for DEN4Δ30 (Whitehead et al. 2003a) and well tolerated and immunogenic in humans (Durbin et al. 2006a). However, introduction of the Δ30 mutation into DENV-2 conferred only a modest level of attenuation (Blaney et al. 2004b) and introduction into DENV-3 failed to attenuate the resulting virus (Blaney et al. 2004a). To create attenuated vaccine candidates for DENV-2 and DENV-3, an additional strategy was developed using molecular techniques to create chimeric viruses. For more than a decade it has been possible to produce chimeric DENV by bringing together the immunogenic structural genes of one DENV and the attenuated non-structural genes of another DENV. The generation of chimeric DENV was first reported by scientists at NIAID in 1991 (Bray and Lai 1991) shortly after their recovery of DENV-4 from a full-length cDNA clone (Lai et al. 1991). As a platform to create attenuated vaccine candidates, DEN4Δ30 was used as the genetic acceptor for the wild-type structural genes of DENV-1, -2 and -3. The Δ30 mutation has been shown to contribute to the attenuation observed for chimeric viruses DEN2/4Δ30 and DENV3/4Δ30 (Blaney et al. 2004a; Whitehead et al. 2003b) and it is clear that the level of attenuation observed for the acceptor virus is maintained in the chimeric virus without the need for attenuating mutations in the structural genes. Tetravalent formulations containing these viruses along with DEN1Δ30 and DEN4Δ30 have been tested in monkeys and shown to be attenuated and elicit balanced antibody responses (Blaney et al. 2005). Vaccine candidate DEN2/4Δ30 has been shown to be safe and immunogenic in adult volunteers at a dose of 10^3 pfu (Durbin et al. 2006b). Preliminary results of the Phase I evaluation of DEN3/4Δ30 have shown the vaccine candidate to be safe but to have an unacceptably low level of infectivity at doses of 10^3 or 10^5 pfu. Alternate DENV-3 vaccine candidates have been generated using full-length DENV-3 in which the 3'-UTR contains two deletion mutations (Δ30/31), or in which the entire 3'-UTR has been replaced with that derived from DEN4Δ30 (Blaney et al. 2008). Phase I evaluation of these new vaccine candidates has been initiated.

Similar to the strategy described above, a live tetravalent chimeric DENV vaccine has been developed utilizing the yellow fever 17D vaccine virus as the genetic background. Four viruses were created in which the prM and E proteins of each DENV1-4 replace those of the yellow fever 17D virus (ChimeriVax-DENV1 – DENV4) (Guirakhoo et al. 2001, 2000). The prME genes used in the chimeric viruses were from virus strains isolated from dengue cases. Although the YF/DENV-2 vaccine virus component appeared immunodominant in early tetravalent formulations, modification of the doses of the individual chimeric viruses in the tetravalent formulation was able to overcome this phenomenon in monkeys (Guirakhoo et al. 2000, 2004b). All 24 monkeys inoculated with one of four different formulations became viremic with mean peak titers ranging from 2.1 $\log_{10}$ –2.6 $\log_{10}$. These titers were similar to those induced by the yellow fever 17D vaccine virus and significantly lower than those induced by the wild-type viruses. Neutralizing antibody titers induced by the tetravalent formulation were similar to those induced by monovalent YF/DENV and the vaccine was highly protective against challenge (Guirakhoo et al. 2004a). Only two monkeys had detectable DENV viremia upon challenge; one with DENV-1 and one with DENV-4. The YF/DENV2 vaccine has been evaluated in healthy adult volunteers (Guirakhoo et al. 2006). The vaccine was found to be safe and highly immunogenic at both low dose (10^3 pfu) and high dose (10^5 pfu) and in both yellow fever immune and nonimmune individuals. The most frequent adverse events associated with the vaccine were headache, myalgia and fatigue. Although a higher percentage of YF/DENV2 vaccinees were viremic than those who received yellow fever vaccine, the peak level of viremia induced by the vaccine was lower than that induced by the yellow fever 17D vaccine. The YF/DENV2 was highly immunogenic in all vaccinees, all of whom seroconverted to DENV-2 and remained seropositive to DENV-2 at 6 and 12 months postvaccination. Cross-reactivity to the other DENV was minimal in YF-naïve individuals but did increase in those with preexisting antibody to YF. This technology has been licensed by Aventis Pasteur (now Sanofi Pasteur) and the YF/DENV1 – 4 tetravalent formulation is currently being evaluated as a three-dose vaccine in Phase 2 studies in dengue endemic regions.

A third chimeric strategy being developed for a tetravalent vaccine is based on the DENV-2 PDK-53 virus (Huang et al. 2000, 2003). The major genetic determinants of attenuation of the DENV-2 PDK-53 virus are located in the nonstructural and untranslated regions of the genome (Butrapet et al. 2000) and the E protein is authentic with the exception of one amino acid change. Recombinant viruses were created in which the prME genes of wildtype DENV-1, -3 and -4 replace those of DENV-2. These viruses were shown to be immunogenic and protective in the mouse model (Huang et al. 2003). Further evaluation of the tetravalent formulation in nonhuman primates and eventual clinical trials are planned.

3 Inactivated Dengue Vaccines

Whole virus inactivated dengue vaccines have three perceived advantages over live attenuated dengue vaccines. These vaccines cannot revert to more virulent viruses, because they do not replicate, they will not interfere with each other in a tetravalent formulation and they can be given to persons who may be immunocompromised. However, these vaccines are generally more expensive to produce, require multiple doses and do not induce the broad or long-lived immune response of live vaccines. The earliest attempts to develop inactivated DENV vaccines were reported more than 75 years ago (Blanc and Caminopetros 1929; Simmons et al. 1931). Subcutaneous vaccination of volunteers with saline suspensions of either filtrates of infected mosquitoes or dried blood that had been collected from dengue-infected patients failed to protect volunteers from subsequent DENV challenge (Simmons et al. 1931). Blanc and Caminopetros reported that DENV in blood was killed when mixed with 1/15th volume of ox bile (Simmons et al. 1931). Although volunteers inoculated with this mixture were not protected against DENV challenge, volunteers who were given a bile-virus mixture of 1:20 were protected from DENV challenge 3 weeks later (Blanc and Caminopetros 1929). The bile-virus mixture of 1:20 did not fully inactivate the virus but may have instead attenuated it thereby enabling the crude vaccine to elicit at least a short-term protective effect.

A purified, formalin-inactivated DENV-2 vaccine (PIV) developed by the U.S. Army was immunogenic in animal models (Putnak et al. 1996a, b). This vaccine was prepared from the DENV-2 strain S16803 and propagated in certified Vero cells. The virus was concentrated by ultracentrifugation and purified over a sucrose gradient resulting in a very high titer virus preparation of approximately 9 $\log_{10}$ pfu per mL. The virus was then inactivated with formalin at final concentration of 0.05% for 10 days at 22°C. The PIV induced neutralizing antibody responses in all macaques after two doses of vaccine and was highly protective against challenge, most notably in those macaques that received higher doses of the PIV or PIV given with one of several adjuvants (Putnak et al. 2005; 1996a).

4 Recombinant Subunit Protein Vaccines

Although live DENV vaccine candidates have been shown to induce protective immunity in nonhuman primate models, developing a live attenuated vaccine with the proper balance of immunogenicity and reactogenicity in humans has been difficult (Edelman et al. 2003; Kitchener et al. 2003; Sun et al. 2003). Recombinant subunit protein vaccines are being evaluated as alternative vaccine strategies to avoid some of these issues. Recombinant DENV proteins can be expressed in baculovirus, yeast, *Escherichia coli*, vaccinia virus and mammalian cells and then purified for use as nonreplicating subunit vaccines. The E protein contains the major antigenic epitopes of the DENV but must be folded in its proper conformation to elicit neutralizing antibodies. This generally requires the coexpression of prM (Allison et al. 1995; Guirakhoo et al. 1992). Full-length E protein, without the coexpression of prM, is targeted intracellulary and is not secreted, failing to induce neutralizing antibody (Feighny et al. 1994; Fonseca et al. 1994). However, when prM and E are expressed together, the integrity of the neutralizing eptiopes is maintained. For this reason, the E protein coexpressed with prM has been most extensively studied in subunit protein vaccines and the baculovirus expression system has been most widely utilized for the expression of these proteins (Bielefeldt-Ohmann et al. 1997; Delenda et al. 1994a; b; Deubel et al. 1991; Eckels et al. 1994; Feighny et al. 1994; Kelly et al. 2000; Putnak et al. 1991; Staropoli et al. 1996, 1997). Removal of the carboxy-terminal end of the E protein containing the membrane anchor site was demonstrated to improve the secretion of the protein and consequently, its immunogenicity (Delenda et al. 1994a, b; Deubel et al. 1991; Men et al. 1991; Staropoli et al. 1996). Purification of a recombinant DENV-2 E protein was improved by fusion of six histidine residues in place of the 100 carboxy-terminal amino acids (Staropoli et al. 1997). The baculovirus-infected cell supernatant was then purified by metal affinity chromatography. The vaccine, when given with aluminum hydroxide as an adjuvant, induced neutralizing antibody and was highly protective in mice. Coexpression of prM and E can induce the formation of viral-like particles (VLPs). Flavivirus VLPs expressed from baculovirus, yeast, or mammalian cells are quite immunogenic, inducing both neutralizing antibody and partial or full protection in the mouse model from wild-type DENV challenge (Kelly et al. 2000; Konishi and Fujii 2002; Konishi et al. 1992; Sugrue et al. 1997).

A recombinant subunit DENV-2 E protein comprised of the N-terminal 80% of E and prM (r80E) expressed in *Drosophilae* cells and given with one of five different adjuvants at day 0 and 30 was evaluated in rhesus macaques (Putnak et al. 2005). Control groups included animals immunized with saline, purified inactivated DENV-2 given with adjuvants, or a single dose of a live attenuated DENV-2 vaccine candidate. Although the live attenuated vaccine candidate provided the best overall protective efficacy against wild-type challenge, the subunit protein vaccines when given with adjuvant induced robust neutralizing antibody titers and provided a high degree of protection against wild-type challenge. Higher levels of neutralizing antibody correlated with protection at challenge. Further evaluation of r80E formulated with different adjuvants is ongoing.

Recombinant fusion proteins comprised of the DENV-2 E protein fused to either the Maltose Binding Protein (MBP) of *E. coli*, the Staphylococcal A protein, or the meningococcal P64K protein were expressed in *E. coli* (Hermida et al. 2004; Simmons et al. 1998; Srivastava et al. 1995). The fusion proteins induced high levels of neutralizing antibody and protection against DENV challenge in the mouse model. When the MBP was cleaved from the E protein, immunogenicity of the protein was greatly decreased. A hybrid protein consisting of the carboxy-truncated DENV-2 E fused in frame with the hepatitis B surface antigen was expressed in *Pichia pastoris*. Despite its ability to form stable VLPs and induce antibody, the antibodies induced were not capable of neutralizing DENV-2 (Bisht et al. 2002). A recombinant hybrid DENV-2/DENV-3 protein comprised of the N-terminal two-thirds of DENV-2 and truncated carboxy-terminal one-third of DENV-3 was expressed in the baculovirus system (Bielefeldt-Ohmann et al. 1997). Although the protein induced virus-specific antibodies to both DENV-2 and DENV-3, it induced neutralizing antibody in mice to DENV-2 only. The protein did induce a strong cross-reactive T cell response. Protective efficacy was not evaluated.

Studies in mice have demonstrated the ability of these proteins to induce neutralizing antibody and to fully or partially protect against challenge with wild-type DENV (Delenda et al. 1994a; Men et al. 1991; Putnak et al. 1991; Simmons et al. 1998; Srivastava et al. 1995). The addition of aluminum hydroxide was shown to increase the amount of neutralizing antibody induced by recombinant DENV-2 proteins, if not the protective efficacy of the vaccines in mice (Kelly et al. 2000; Staropoli et al. 1997). Despite generally promising results in mice, the subunit protein vaccines described above were inferior to live DENV vaccines when tested in nonhuman primates, providing only partial protection against challenge with wild-type DENV (Eckels et al. 1994; Guzman et al. 2003; Velzing et al. 1999). No subunit protein DENV vaccine has entered clinical trial to date.

5 DNA Vaccines

DNA vaccines are thought to have several advantages over traditional inactivated whole virus vaccines, recombinant protein vaccines, subunit protein vaccines and, to some extent, live attenuated vaccines (Whalen 1996). DNA is stable for long periods of time and is resistant to extremes of temperature, overcoming cold-chain restrictions. Because the proteins produced by DNA vaccines are translated and processed within the host cell, they are able to induce class I MHC-dependent immune responses. In addition, DNA vaccines cause less reactogenicity than live vaccines yet are able to induce long-lived humoral and cellular immunity (Rhodes et al. 1994). Kochel et al. at the Naval Medical Research Center (NMRC) evaluated the immunogenicity of the plasmid expression vector VR1 012 containing the PrM gene and 92% of the E gene of DENV-2 New Guinea C (NGC) in Balb/C mice (Kochel et al. 1997). Neutralizing antibody was induced in all vaccinated mice and a higher survival rate was observed in vaccinated mice following challenge with wild-type DENV-2 virus (Kochel et al. 1997; Porter et al. 1998). Neutralizing antibody titers in mice were significantly increased when the DNA construct was engineered such that the expressed DENV-2 E protein was fused to lysosome-associated membrane protein (LAMP) to enhance trafficking of the protein to the lysosome for improved MHC class II expression (Raviprakash et al. 2001). Subsequently, four different DENV-1 DNA constructs were evaluated in mice and it was determined that the plasmid expressing the prM and full-length DENV-1 E protein (DENV-1ME) elicited the highest antibody response (Raviprakash et al. 2000a). DEN1ME was further studied in rhesus macaques and *Aotus* monkeys to evaluate its protective efficacy against challenge (Kochel et al. 2000; Raviprakash et al. 2000b). In these studies, the DENV-1ME vaccine only partially protected the monkeys against challenge with wild-type DENV-1 Western Pacific 74. Studies were then performed by the NMRC group in *Aotus* monkeys to evaluate whether the co-administration of DENV-1ME with plasmids expressing multiple copies of human immunostimulatory sequences (ISS) or *Aotus* granulocyte macrophage colony stimulating factor (GM-CSF) could improve the immunogenicity and protective efficacy of the vaccine (Raviprakash et al. 2003). Although the co-administration of plasmids expressing immunomodulators did not significantly increase the DENV-1 neutralizing antibody titers, 100% of the animals were protected against wild-type DENV-1 challenge 6 months after vaccination, compared with the 33% protection observed in *Aotus* monkeys vaccinated with DENV-1ME alone in the earlier study. The DENV-1ME DNA vaccine is currently being evaluated in healthy human adult volunteers in the United States.

Konishi and colleagues successfully immunized mice with a tetravalent dengue DNA vaccine consisting of four plasmids expressing the prM/E of each DENV serotype (Konishi et al. 2006). The tetravalent DNA vaccine induced comparable antibody titers to those induced by the individual plasmids when given as monovalent vaccines, demonstrating that DNA vaccines may overcome the hurdle of viral interference that was noted with some live attenuated DENV vaccines (see above) (Konishi et al. 2006). Apt and colleagues at the NRMC have utilized DNA "shuffling" and screening to develop a tetravalent DNA vaccine expressing a chimeric DENV envelope antigen capable of inducing neutralizing antibody to all four DENV serotypes (Apt et al. 2006; Raviprakash et al. 2006). Although the constructs evaluated were capable of inducing neutralizing antibody against all four serotypes, vaccinated macaques were not protected against DENV-2 challenge (Raviprakash et al. 2006). The authors surmise that the DENV-2 neutralizing antibody titers induced by the DNA shuffle vaccines were not sufficient to protect against DENV-2 infection and that strategies to enhance this immune response should be evaluated.

The protective efficacy of a DNA vaccine expressing the nonstructural protein NS1 has been evaluated in mouse models (Costa et al. 2007; Wu et al. 2003). Because NS1 antibody is nonneutralizing and does not bind to the virus, it is proposed that an NSI vaccine would eliminate the risk of antibody dependent enhancement in vaccine recipients, even as antibody titers waned over time. Mice immunized with the NS1 vaccine were partially protected from wild-type DENV challenge (Costa et al. 2007; Wu et al. 2003) and this was enhanced by administration of a plasmid expressing IL-12 (Wu et al. 2003).

6   Virus-Vectored DENV Vaccines

Recombinant poxviruses and adenoviruses expressing foreign proteins have been demonstrated to induce strong humoral and cellular responses in humans against various pathogens (Catanzaro et al. 2006; Liniger et al. 2007; Moss 1996). These viruses can infect cells and express their proteins de novo within the cell. The antigens are then naturally processed, glycosylated and associated with the cell membrane. In addition, because of intracellular translation and processing of the gene products, MHC class I dependent immune responses are induced. Despite these benefits, early studies of recombinant vaccinia viruses expressing the structural proteins of DENV-2 or DENV-4 were disappointing (Bray et al. 1989; Deubel et al. 1988). These constructs expressed prM and a full-length E protein but failed to induce neutralizing antibody and failed to protect monkeys from wild-type challenge. Vaccinia recombinants expressing truncated DENV E proteins were capable of inducing neutralizing antibody in mice and in rhesus macaques and afforded protection against wild-type challenge in both models (Men et al. 2000, 1991). It was hypothesized that the poor immunogenicity of the previous constructs was due to intracellular targeting of the E protein (see recombinant subunit protein vaccines, above).

A replication-defective recombinant adenovirus expressing the prM and carboxy-truncated E proteins was constructed by inserting the prME gene of DEN2 NGC into the early region 1 of the adenovirus genome (Jaiswal et al. 2003). Mice were immunized at 0, 1 and 2 months with 10^7 pfu of the recombinant adenovirus (rAd). After the third injection, neutralizing antibody titers induced by the rAd against DENV-2 were comparable to those induced by a purified inactivated DENV-2 vaccine. Two complex rAd vectors were developed that express 4 the prM and E proteins of DENV-1 and DENV-2 [cAdVaxD(1-2)] or the prM and E of DENV-3 and DENV-4 [cAdVaxD(3-4)] (Holman et al. 2007; Raja et al. 2007). Both constructs induced neutralizing antibody against all four DENV serotypes and a cellular immune response was detected 4–10 weeks following primary vaccination (Holman et al. 2007). The authors subsequently evaluated the two recombinant adenovirus vectors given together as a two-dose tetravalent DENV vaccine in rhesus monkeys and demonstrated significant protection against DENV challenge (Raviprakash et al. 2008).

Venezuelan equine encephalitis virus (VEE) has been utilized as a vector for a recombinant of DENV-1 vaccine (Chen et al. 2007). A recombinant VEE expressing the prM and full-length E of DENV-1 as a virus replicon particle (VRP) was evaluated in cynomologous macaques. The DENV-1 VRP was given at 0, 1 and 4 months and was compared with the naked DENV-1 DNA vaccine described above (Raviprakash et al. 2000b). The DENV-1 VRP induced comparable neutralizing antibody titers to three doses of the naked DNA vaccine in immunized macaques and both vaccines were partially protective against wild-type DENV-1 challenge (Chen et al. 2007). When the VRP was used in a heterologous prime-boost regimen, superior results were obtained. The DENV-1 DNA vaccine was given at times 0 and 1 month, followed by the DENV-1 VRP at 4 months. This regimen induced complete protection against DENV-1 challenge in all vaccinated macaques more than 4 months after the last vaccination.

7   Conclusion

In conclusion, live attenuated DENV vaccines have been most extensively evaluated in clinical trials and are furthest along in the development pipeline. The insufficient potency of dengue DNA and subunit protein vaccines in nonhuman primate models and the concern that preexisting and acquired immunity to poxvirus and adenovirus vectors may diminish their effectiveness as vaccines are current obstacles that must be overcome. Novel adjuvants and prime-boost strategies have had some success in improving the performance of these candidates and should be further evaluated (Chen et al. 2007; Imoto and Konishi, 2007; Putnak et al. 2005; Wu et al. 2003; Yang et al. 2003). It is anticipated that clinical trials evaluating novel recombinant subunit protein, DNA and vectored vaccines will be initiated in the coming years, either alone or as part of a prime-boost strategy. In addition, clinical trials of several new live attenuated DENV vaccine candidates are planned.

References

Allison SL, Stadler K, Mandl CW, Kunz C, Heinz FX (1995) J Virol 69:5816–5820

Apt D, Raviprakash K, Brinkman A, Semyonov A, Yang S, Skinner C, Diehl L, Lyons R, Porter K, Punnonen J (2006) Vaccine 24:335–344

Bancroft WH, Scott RM, Eckels KH, Hoke CH Jr, Simms TE, Jesrani KD, Summers PL, Dubois DR, Tsoulos D, Russell PK (1984) J Infect Dis 149:1005–1010

Bancroft WH, Top FH Jr, Eckels KH, Anderson JH Jr, McCown JM, Russell PK (1981) Infect Immun 31:698–703

Bhamarapravati N, Sutee Y (2000) Vaccine 18:44–47

Bhamarapravati N, Yoksan S (1989) Lancet 1:1077

Bhamarapravati N, Yoksan S (1997) Live attenuated tetravalent dengue vaccine. In: Gubler DJ, Kuno G (eds) Dengue and dengue hemorrhagic fever. CABI Publishing, New York, pp 367–377

Bhamarapravati N, Yoksan S, Chayaniyayothin T, Angsubphakorn S, Bunyaratvej A (1987) Bull WHO 65:189–195

Bielefeldt-Ohmann H, Beasley DW, Fitzpatrick DR, Aaskov JG (1997) J Gen Virol 78:2723–2733

Bisht H, Chugh DA, Raje M, Swaminathan SS, Khanna N (2002) J Biotechnol 99:97–110

Blanc G, Caminopetros J (1929) Bull Acad Med 102:37–40

Blaney JE Jr, Hanson CT, Firestone CY, Hanley KA, Murphy BR, Whitehead SS (2004a) Am J Trop Med Hyg 71:811–821

Blaney JE Jr, Hanson CT, Hanley KA, Murphy BR, Whitehead SS (2004b) BMC Infect Dis 4:39

Blaney JE Jr, Matro JM, Murphy BR, Whitehead SS (2005) J Virol 79:5516–5528

Blaney JE Jr, Sathe NS, Goddard L, Hanson CT, Romero TA, Hanley KA, Murphy BR, Whitehead SS (2008) Vaccine 26:817–828

Bray M, Lai CJ (1991) Proc Natl Acad Sci USA 88:10342–10346

Bray M, Zhao BT, Markoff L, Eckels KH, Chanock RM, Lai CJ (1989) J Virol 63:2853–2856

Butrapet S, Huang CY, Pierro DJ, Bhamarapravati N, Gubler DJ, Kinney RM (2000) J Virol 74:3011–3019

Catanzaro AT, Koup RA, Roederer M, Bailer RT, Enama ME, Moodie Z, Gu L, Martin JE, Novik L, Chakrabarti BK, Butman BT, Gall JG, King CR, Andrews CA, Sheets R, Gomez PL, Mascola JR, Nabel GJ, Graham BS (2006) J Infect Dis 194:1638–1649

Chen L, Ewing D, Subramanian H, Block K, Rayner J, Alterson KD, Sedegah M, Hayes C, Porter K, Raviprakash K (2007) J Virol 81:11634–11639

Costa SM, Azevedo AS, Paes MV, Sarges FS, Freire MS, Alves AM (2007) Virology 358:413–423

Delenda C, Frenkiel MP, Deubel V (1994a) Arch Virol 139:197–207

Delenda C, Staropoli I, Frenkiel MP, Cabanie L, Deubel V (1994b) J Gen Virol 75:1569–1578

Deubel V, Bordier M, Megret F, Gentry MK, Schlesinger JJ, Girard M (1991) Virology 180:442–447

Deubel V, Kinney RM, Esposito JJ, Cropp CB, Vorndam AV, Monath TP, Trent DW (1988) J Gen Virol 69:1921–1929

Durbin AP, Karron RA, Sun W, Vaughn DW, Reynolds MJ, Perreault JR, Thumar B, Men R, Lai CJ, Elkins WR, Chanock RM, Murphy BR, Whitehead SS (2001) Am J Trop Med Hyg 65:405–413

Durbin AP, McArthur J, Marron JA, Blaney JE Jr, Thumar B, Wanionek K, Murphy BR, Whitehead SS (2006a) Hum Vaccin 2:167–173

Durbin AP, McArthur JH, Marron JA, Blaney JE, Thumar B, Wanionek K, Murphy BR, White-head SS (2006b) Hum Vaccin 2:255–260

Durbin AP, Whitehead SS, McArthur J, Perreault JR, Blaney JE Jr, Thumar B, Murphy BR, Karron RA (2005) J Infect Dis 191:710–718

Eckels KH, Dubois DR, Summers PL, Schlesinger JJ, Shelly M, Cohen S, Zhang YM, Lai CJ, Kurane I, Rothman A, Hasty S, Howard B (1994) Am J Trop Med Hyg 50:472–478

Eckels KH, Scott RM, Bancroft WH, Brown J, Dubois DR, Summers PL, Russell PK, Halstead SB (1984) Am J Trop Med Hyg 33:684–689

Edelman R, Tacket CO, Wasserman SS, Vaughn DW, Eckels KH, Dubois DR, Summers PL, Hoke CH (1994) J Infect Dis 170:1448–1455

Edelman R, Wasserman SS, Bodison SA, Putnak RJ, Eckels KH, Tang D, Kanesa-Thasan N, Vaughn DW, Innis BL, Sun W (2003) Am J Trop Med Hyg 69:48–60

Feighny R, Burrous J, Putnak R (1994) Am J Trop Med Hyg 50:322–328

Fonseca BA, Pincus S, Shope RE, Paoletti E, Mason PW (1994) Vaccine 12:279–285

Fujita N, Oda K, Yasui Y, Hotta S (1969) Kobe J Med Sci 15:163–180

Guirakhoo F, Arroyo J, Pugachev KV, Miller C, Zhang ZX, Weltzin R, Georgakopoulos K, Catalan J, Ocran S, Soike K, Ratterree M, Monath TP (2001) J Virol 75:7290–7304

Guirakhoo F, Bolin RA, Roehrig JT (1992) Virology 191:921–931

Guirakhoo F, Kitchener S, Morrison D, Forrat R, McCarthy K, Nichols R, Yoksan S, Duan X, Ermak TH, Kanesa-Thasan N, Bedford P, Lang J, Quentin-Millet MJ, Monath TP (2006) Hum Vaccin 2:60–67

Guirakhoo F, Pugachev K, Zhang Z, Myers G, Levenbook I, Draper K, Lang J, Ocran S, Mitchell F, Parsons M, Brown N, Brandler S, Fournier C, Barrere B, Rizvi F, Travassos A, Nichols R, Trent D, Monath T (2004a) J Virol 78:4761–4775

Guirakhoo F, Weltzin R, Chambers TJ, Zhang Z, Soike K, Ratterree M, Arroyo J, Georgakopoulos K, Catalan J, Monath TP (2000) J Virol 74:5477–5485

Guirakhoo F, Zhang Z, Myers G, Johnson BW, Pugachev K, Nichols R, Brown N, Levenbook I, Draper K, Cyrek S, Lang J, Fournier C, Barrere B, Delagrave S, Monath TP (2004b) J Virol 78:9998–10008

Guzman MG, Rodriguez R, Hermida L, Alvarez M, Lazo L, Mune M, Rosario D, Valdes K, Vazquez S, Martinez R, Serrano T, Paez J, Espinosa R, Pumariega T, Guillen G (2003) Am J Trop Med Hyg 69:129–134

Hermida L, Rodriguez R, Lazo L, Silva R, Zulueta A, Chinea G, Lopez C, Guzman MG, Guillen G (2004) A dengue-2 Envelope fragment inserted within the structure of the P64k meningococcal protein carrier enables a functional immune response against the virus in mice. J Virol Methods 115:41–49

Holman DH, Wang D, Raviprakash K, Raja NU, Luo M, Zhang J, Porter KR, Dong JY (2007) Clin Vaccine Immunol 14:182–189

Hotta S (1952) J Infect Dis 90:1–9

Hotta S (1969) In: Sanders M (ed) Dengue and Related Hemorrhagic Diseases. Warren H. Green Inc, St. Louis, p 166

Huang CY, Butrapet S, Pierro DJ, Chang GJ, Hunt AR, Bhamarapravati N, Gubler DJ, Kinney RM (2000) J Virol 74:3020–3028

Huang CY, Butrapet S, Tsuchiya KR, Bhamarapravati N, Gubler DJ, Kinney RM (2003) J Virol 77:11436–11447

Imoto J, Konishi E (2007) Vaccine 25:1076–1084

Innis BL, Eckels KH, Kraiselburd E, Dubois DR, Meadors GF, Gubler DJ, Burke DS, Bancroft WH (1988) J Infect Dis 158:876–880

Jaiswal S, Khanna N, Swaminathan S (2003) J Virol 77:12907–12913

Kanesa-Thasan N, Edelman R, Tacket CO, Wasserman SS, Vaughn DW, Coster TS, Kim-Ahn GJ, Dubois DR, Putnak JR, King A, Summers PL, Innis BL, Eckels KH, Hoke CH Jr (2003) Am J Trop Med Hyg 69:17–23

Kanesa-thasan N, Sun W, Kim-Ahn G, Van Albert S, Putnak JR, King A, Raengsakulsrach B, Christ-Schmidt H, Gilson K, Zahradnik JM, Vaughn DW, Innis BL, Saluzzo JF, Hoke CH Jr (2001) Vaccine 19:3179–3188

Kelly EP, Greene JJ, King AD, Innis BL (2000) Vaccine 18:2549–2559

Kitchener S, Baade L, Brennan L (2003) J Travel Med 10:50–51

Kitchener S, Nissen M, Nasveld P, Forrat R, Yoksan S, Lang J, Saluzzo JF (2006) Vaccine 24:1238–1241

Kochel T, Wu SJ, Raviprakash K, Hobart P, Hoffman S, Porter K, Hayes C (1997) Vaccine 15:547–552

Kochel TJ, Raviprakash K, Hayes CG, Watts DM, Russell KL, Gozalo AS, Phillips IA, Ewing DF, Murphy GS, Porter KR (2000) Vaccine 18:3166–3173

Konishi E, Fujii A (2002) Vaccine 20:1058–1067

Konishi E, Kosugi S, Imoto J (2006) Vaccine 24:2200–2207

Konishi E, Pincus S, Paoletti E, Shope RE, Burrage T, Mason PW (1992) Virology 188:714–720

Lai CJ, Zhao BT, Hori H, Bray M (1991) Proc Natl Acad Sci USA 88:5139–5143

Liniger M, Zuniga A, Naim HY (2007) Expert Rev Vaccines 6:255–266

McKee KT Jr, Bancroft WH, Eckels KH, Redfield RR, Summers PL, Russell PK (1987) Am J Trop Med Hyg 36:435–442

Men R, Bray M, Clark D, Chanock RM, Lai CJ (1996) J Virol 70:3930–3937

Men R, Wyatt L, Tokimatsu I, Arakaki S, Shameem G, Elkins R, Chanock R, Moss B, Lai CJ (2000) Vaccine 18:3113–3122

Men RH, Bray M, Lai CJ (1991) J Virol 65:1400–1407

Moss B (1996) Proc Natl Acad Sci USA 93:11341–11348

Porter KR, Kochel TJ, Wu SJ, Raviprakash K, Phillips I, Hayes CG (1998) Arch Virol 143:997–1003

Putnak JR, Coller BA, Voss G, Vaughn DW, Clements D, Peters I, Bignami G, Houng HS, Chen RC, Barvir DA, Seriwatana J, Cayphas S, Garcon N, Gheysen D, Kanesa-Thasan N, McDonell M, Humphreys T, Eckels KH, Prieels JP, Innis BL (2005) Vaccine 23:4442–4452

Putnak R, Barvir DA, Burrous JM, Dubois DR, D'Andrea VM, Hoke CH, Sadoff JC, Eckels KH (1996a) J Infect Dis 174:1176–1184

Putnak R, Cassidy K, Conforti N, Lee R, Sollazzo D, Truong T, Ing E, Dubois D, Sparkuhl J, Gastle W, Hoke CH Jr (1996b) Am J Trop Med Hyg 55:504–510

Putnak R, Feighny R, Burrous J, Cochran M, Hackett C, Smith G, Hoke CH Jr (1991) Am J Trop Med Hyg 45:159–167

Raja NU, Holman DH, Wang D, Raviprakash K, Juompan LY, Deitz SB, Luo M, Zhang J, Porter KR, Dong JY (2007) Am J Trop Med Hyg 76:743–751

Raviprakash K, Apt D, Brinkman A, Skinner C, Yang S, Dawes G, Ewing D, Wu SJ, Bass S, Punnonen J, Porter K (2006) Virology 353:166–173

Raviprakash K, Ewing D, Simmons M, Porter KR, Jones TR, Hayes CG, Stout R, Murphy GS (2003) Virology 315:345–352

Raviprakash K, Kochel TJ, Ewing D, Simmons M, Phillips I, Hayes CG, Porter KR (2000a) Vaccine 18:2426–2434

Raviprakash K, Marques E, Ewing D, Lu Y, Phillips I, Porter KR, Kochel TJ, August TJ, Hayes CG, Murphy GS (2001) Virology 290:74–82

Raviprakash K, Porter KR, Kochel TJ, Ewing D, Simmons M, Phillips I, Murphy GS, Weiss WR, Hayes CG (2000b) J Gen Virol 81(Pt 7):1659–1667

Raviprakash K, Wang D, Ewing D, Holman DH, Block K, Woraratanadharm J, Chen L, Hayes C, Dong JY, Porter K (2008) A Tetravalent Dengue Vaccine Based on a Complex Adenovirus Vector Provides Significant Protection in Rhesus Monkeys against All Four Serotypes of Dengue Virus. J Virol 81(14):6927–6934

Rhodes GH, Abai AM, Margalith M, Kuwahara-Rundell A, Morrow J, Parker SE, Dwarki VJ (1994) Dev Biol Stand 82:229–236

Sabchareon A, Lang J, Chanthavanich P, Yoksan S, Forrat R, Attanath P, Sirivichayakul C, Pengsaa K, Pojjaroen-Anant C, Chambonneau L, Saluzzo JF, Bhamarapravati N (2004) Pediatr Infect Dis J 23:99–109

Sabchareon A, Lang J, Chanthavanich P, Yoksan S, Forrat R, Attanath P, Sirivichayakul C, Pengsaa K, Pojjaroen-Anant C, Chokejindachai W, Jagsudee A, Saluzzo JF, Bhamarapravati N (2002) Am J Trop Med Hyg 66:264–272

Sabin A (1952) Am J Trop Med Hyg 1:30–50

Sabin AB, Schlesinger RW (1945) Science 101:640–642

Schlesinger RW, Gordon I, Frankel JW, Winter JW, Patterson PR, Dorrance WR (1956) J Immunol 77:352–364

Scott RM, Eckels KH, Bancroft WH, Summers PL, McCown JM, Anderson JH, Russell PK (1983) J Infect Dis 148:1055–1060

Simasathien S, Thomas SJ, Watanaveeradej V, Nisalak A, Barberousse C, Innis BL, Sun W, Putnak JR, Eckels KH, Hutagalung Y, Gibbons RV, Zhang C, De La Barrera R, Jarman RG, Chawachalasai W, Mammen MP Jr (2008) Am J Trop Med Hyg 78:426–433

Simmons JS, St John JH, Reynolds FHK (1931) Philipp J Sci 44:1–252

Simmons M, Nelson WM, Wu SJ, Hayes CG (1998) Am J Trop Med Hyg 58:655–662

Srivastava AK, Putnak JR, Warren RL, Hoke CH Jr (1995) Vaccine 13:1251–1258

Staropoli I, Clement JM, Frenkiel MP, Hofnung M, Deubel V (1996) J Virol Methods 56:179–189

Staropoli I, Frenkiel MP, Megret F, Deubel V (1997) Vaccine 15:1946–1954

Sugrue RJ, Fu J, Howe J, Chan YC (1997) J Gen Virol 78:1861–1866

Sun W, Edelman R, Kanesa-Thasan N, Eckels KH, Putnak JR, King AD, Houng HS, Tang D, Scherer JM, Hoke CH Jr, Innis BL (2003) Am J Trop Med Hyg 69:24–31

Vaughn DW, Hoke CH Jr, Yoksan S, LaChance R, Innis BL, Rice RM, Bhamarapravati N (1996) Vaccine 14:329–336

Velzing J, Groen J, Drouet MT, van Amerongen G, Copra C, Osterhaus AD, Deubel V (1999) Vaccine 17:1312–1320

Whalen RG (1996) Emerg Infect Dis 2:168–175

Whitehead SS, Blaney JE, Durbin AP, Murphy BR (2007) Nat Rev Microbiol 5:518–528

Whitehead SS, Falgout B, Hanley KA, Blaney JE Jr, Markoff L, Murphy BR (2003a) J Virol 77:1653–1657

Whitehead SS, Hanley KA, Blaney JE Jr, Gilmore LE, Elkins WR, Murphy BR (2003b) Vaccine 21:4307–4316

Wisseman CL Jr, Sweet BH, Rosenzweig EC, Eylar OR (1963) Am J Trop Med Hyg 12:620–623

Wu SF, Liao CL, Lin YL, Yeh CT, Chen LK, Huang YF, Chou HY, Huang JL, Shaio MF, Sytwu HK (2003) Vaccine 21:3919–3929

Yang ZY, Wyatt LS, Kong WP, Moodie Z, Moss B, Nabel GJ (2003) J Virol 77:799–803

Targeted Mutagenesis as a Rational Approach to Dengue Virus Vaccine Development

Joseph E. Blaney Jr., Anna P. Durbin, Brian R. Murphy, and Stephen S. Whitehead

Contents

Abstract The recombinant dengue virus type 4 (rDEN4) vaccine candidate, rDEN4Δ30, was found to be highly infectious, immunogenic and safe in human volunteers. At the highest dose (10^5 PFU) evaluated in volunteers, 25% of the vaccinees had mild elevations in liver enzymes that were rarely seen at lower doses. Here, we describe the generation and selection of additional mutations that were introduced into rDEN4Δ30 to further attenuate the virus in animal models and ultimately human vaccinees. Based on the elevated liver enzymes associated with the 10^5 PFU dose of rDEN4Δ30 and the known involvement of liver infection in dengue virus pathogenesis, a large panel of mutant viruses was screened for level of replication in the HuH-7 human hepatoma cell line, a

J.E. Blaney Jr., B.R. Murphy and S.S. Whitehead (✉)
Laboratory of Infectious Diseases, National Institute of Allergy and Infectious Diseases, National Institutes of Health, 33 North Drive, Room 3W10A, Bethesda, MD 20892-3203, USA
e-mail: swhitehead@niaid.nih.gov

A.P. Durbin
Department of International Health, Center for Immunization Research, Johns Hopkins Bloomberg School of Public Health, 624 N. Broadway, Rm. 251, Baltimore, MD 21205, USA

A.L. Rothman (ed.), *Dengue Virus*, Current Topics in Microbiology and Immunology 338,   145
DOI 10.1007/978-3-642-02215-9_11, © Springer-Verlag Berlin Heidelberg 2010

surrogate for human liver cells and selected viruses were further analyzed for level of viremia in SCID-HuH-7 mice. It was hypothesized that rDEN4Δ30 derivatives with restricted replication in vitro and in vivo in HuH-7 human liver cells would be restricted in replication in the liver of vaccinees. Two mutations identified by this screen, NS3 4995 and NS5 200,201, were separately introduced into rDEN4Δ30 and found to further attenuate the vaccine candidate for SCID-HuH-7 mice and rhesus monkeys while retaining sufficient immunogenicity in rhesus monkeys to confer protection. In humans, the rDEN4Δ30-200,201 vaccine candidate administered at 10^5 PFU exhibited greatly reduced viremia, high infectivity and lacked liver toxicity while inducing serum neutralizing antibody at a level comparable to that observed in volunteers immunized with rDEN4Δ30. Clinical studies of rDEN4Δ30-4995 are ongoing.

1 Introduction

The four DENV serotypes, DENV-1, DENV-2, DENV-3 and DENV-4, cause considerably more human disease than any other arbovirus. DENV is currently receiving heightened research interest because of the intense spread and increased prevalence of the disease throughout tropical and subtropical regions of the world. In fact, dengue hemorrhagic fever (DHF) cases have now been confirmed in more than 60 countries and DENV is endemic in more than 100 countries including most of Southeast Asia, South America, Central America and the Caribbean and South Pacific regions (Gubler and Meltzer 1999; WHO 1997). Prior to the 1970s, only five Southeast Asian countries had documented DHF (Gubler 1997). Currently a licensed vaccine is not available to confer protection from dengue disease but the development of a tetravalent DENV vaccine has recently received considerable effort with substantial advancement (Whitehead et al. 2007).

Live attenuated tetravalent vaccines for DENV are currently in the most advanced stages of development. Nevertheless, several factors have complicated the development of an effective DENV vaccine including the lack of an animal model for DENV disease and the need to simultaneously protect against all four serotypes. A vaccine must confer protection from each of the DENV serotypes because all four serotypes commonly circulate in endemic regions and secondary infection with a heterologous serotype is associated with increased disease severity by a mechanism termed antibody-dependent enhancement (Halstead 2003). It is imperative that an acceptable DENV vaccine elicits a consistently strong and protective neutralizing antibody response against each of the four serotypes using a tetravalent formulation of antigens. The benefits of live attenuated vaccines for DENV are believed to outweigh the potential risks inherent with live virus vaccines. Live attenuated vaccines can induce potent humoral and cellular immune responses, mimic natural infection and can provide lifelong immunity. In addition, live attenuated vaccines are the most economical option because of the low cost of manufacture in Vero cell tissue culture and the high level of infectivity of the vaccine viruses (Blaney et al. 2006).

2   The rDEN4Δ30 Vaccine Candidate

Conventional methods of serial passage of virus in an animal host (e.g., mouse brain) or in tissue culture were used to develop the first generation live attenuated DENV vaccine candidates (Bhamarapravati and Yoksan 1997; Halstead and Marchette 2003; Sabin 1952). More recently, recombinant DNA methodology and reverse genetics has been utilized to design live attenuated recombinant DENV vaccine candidates (Durbin et al. 2001; Guirakhoo et al. 2006). For recovery of an infectious flavivirus, which contains a single-stranded, positive-sense RNA genome, RNA transcripts are produced from cloned cDNAs in vitro and then delivered into susceptible tissue culture cells by transfection or electroporation (Lai et al. 1991; Rice et al. 1989). This methodology allows recovery of well-defined, genetically homogeneous virus populations with a safe and easily characterized passage history. Site directed mutagenesis of cloned virus cDNAs has provided mutant viruses bearing engineered mutations that can be examined in vivo for their level of attenuation and stability (Blaney et al. 2002; Hanley et al. 2002; Men et al. 1996). Importantly, recombinant DNA methodology can further be used to modify and combine mutations to develop a myriad of well-defined vaccine candidates.

The first reported recombinant DENV to be evaluated in humans was the rDEN4Δ30 vaccine candidate that contains a 30 nucleotide deletion in the $3'$ UTR, removing nucleotides 10,478–10,507 (Durbin et al. 2001, 2005). Preclinical evaluation of rDEN4Δ30 demonstrated that the vaccine candidate had restricted replication in rhesus monkeys as indicated by a shortened duration and lower magnitude of viremia compared to wild type DENV-4 (Durbin et al. 2001; Men et al. 1996). Comparison of DENV-4 and rDEN4Δ30 virus replication in mosquitoes indicated that the Δ30 mutation restricted dissemination from the mid-gut to the head of infected mosquitoes, blocking a necessary step for virus transmission (Troyer et al. 2001). Importantly, the rDEN4Δ30 virus replicates to titers of $> 10^7$ PFU/ml in qualified Vero cells, which are suitable for vaccine production.

For the initial clinical evaluation, 20 adult volunteers were administered a single subcutaneous dose of 10^5 PFU rDEN4Δ30 (Table 1) (Durbin et al. 2001). Serious clinical responses were not observed and the vaccine was well tolerated. The most frequent clinical reactions were transient and asymptomatic including a maculo-papular rash in 50% of volunteers, increased alanine aminotransferase (ALT) levels in 25% of volunteers and neutropenia in 15% of volunteers. Transient elevations in serum liver enzymes are also seen in vaccinees infected with other live attenuated DENV vaccine candidates (Eckels et al. 1984; Edelman et al. 1994; Kanesa-thasan et al. 2001; Vaughn et al. 1996) and the levels observed in vaccinees are generally much lower than those observed in humans experiencing DF or DHF/DSS. Furthermore, the hepatomegaly that is observed in some dengue patients (Kalayanarooj et al. 1997; Kuo et al. 1992; Mohan et al. 2000; Wahid et al. 2000) has not been observed in vaccinees. Hepatotropism is a natural feature of DENV infection of humans and the rDEN4Δ30 vaccine candidate likely exhibited a low level of residual hepatotropism with asymptomatic hepatotoxicity at the high dose of

Table 1 Effect of rDEN4Δ30 vaccine dose on virus replication, immunogenicity and clinical signs in dose cohorts of 20 volunteers

Dose (PFU)	% of volunteers with viremia	Mean peak virus titer ± SE ($\log_{10}$ PFU/ml serum)[a]	Geometric mean serum neutralizing antibody titer[b]	% seroconversion[c]	% of volunteers with indicated clinical sign:				
					Fever	Rash	Headache	Neutropenia[d]	Elevated ALT level[e]
10^5	70	1.6±0.1	399	100	5[f]	50	35	15	25
10^3	35	0.5±0.1	129	95	0	55	35	25	5
10^2	55	0.7±0.1	181	95	5[g]	80	45	20	0
10^1	65	0.6±0.1	355	100	0	75	45	25	0
Placebo[h]	0	<0.5	<10	0	0	0	42	0	0

[a] Calculated for viremic volunteers only. The lower limit of detection is 0.5 $\log_{10}$ PFU/ml serum

[b] Reciprocal plaque reduction (60%) neutralizing antibody titer on day 42

[c] Percent seroconversion defined as a fourfold or greater increase in serum neutralizing antibody level to DENV-4

[d] Neutropenia defined as an absolute neutrophil count of $< 1,500$ cells/mm^3

[e] Elevated ALT level defined as any value above the upper limit of normal (for males, >72 U/L^{-1}; for females, >52 U/L^{-1})

[f] Fever occurred on days 3 and 5; maximum temp was 100.5°F

[g] Fever occurred on day 3; maximum temp was 100.5°F

[h] n = 12

10^5 PFU. The mean peak titer of rDEN4Δ30 virus in serum ($10^{1.6}$ PFU/ml) of vaccinees is much lower than that observed in natural DF and DHF, which can be 10^6 infectious units/ml or greater (Endy et al. 2004; Vaughn et al. 2000). Despite the low levels of viremia observed in vaccinees, the rDEN4Δ30 vaccine candidate was strongly immunogenic. Each vaccinee (viremic and nonviremic) seroconverted to DENV-4 with a geometric mean neutralizing antibody titer (GMT) of 1:399 (seroconversion is defined as a fourfold or greater increase in neutralizing antibody level after immunization). Importantly, *Aedes* mosquitoes were fed on a subset of volunteers around the time of peak of viremia and the vaccine virus was not transmitted from the infected vaccinees to the feeding mosquitoes (Troyer et al. 2001).

While the initial evaluation of rDEN4Δ30 at 10^5 PFU indicated that the vaccine was well-tolerated and safe, two approaches were pursued to decrease the mild reactogenicity. First, a placebo-controlled dose response study was performed to determine the 50% human infectious dose (HID$_{50}$) of the vaccine and to examine if lowering the dose could reduce the rash or the laboratory abnormalities observed in vaccinees (Durbin et al. 2005). The HID$_{50}$ was determined to be < 10 PFU since 95–100% of vaccinees seroconverted at doses of 10^3, 10^2 and 10^1 PFU (Table 1). The frequency of detectable viremia in volunteers was not different among the dose cohorts but mean peak virus titers among viremic volunteers were approximately tenfold lower for the 10^3, 10^2 and 10^1 PFU cohorts in comparison to the 10^5 PFU cohort. In addition, 95–100% of volunteers seroconverted to DENV-4 with consistently high neutralizing antibody titers regardless of vaccination dose. At a dose of 10^3 PFU, only a single volunteer had an elevated serum ALT level and elevated ALT levels were not detected in any volunteer in the 10^2 or 10^1 PFU cohorts, indicating that reduction of dose was effective in decreasing the mild hepatotoxicity of rDEN4Δ30. However, the frequency of rash and neutropenia was not related to vaccine dose and remained similar in each dose cohort. Studies performed by others have also indicated that reducing the vaccine dose may not reliably reduce clinical signs or laboratory abnormalities (Kitchener et al. 2006; Sabchareon et al. 2004, 2002).

We also pursued an additional genetic approach to reduce the mild reactogenicity observed with rDEN4Δ30 by developing further attenuated variants of rDEN4Δ30 by the introduction of additional attenuating mutations. We especially sought mutations that limited replication in vitro or in vivo in HuH-7 cells, our surrogate for human liver cells.

3   Development of Further Attenuated rDEN4Δ30 Vaccine Candidates

3.1   Generation of Mutant rDEN4 Viruses

The identification of mutations that attenuate DENV has been a major focus of our strategy for the development of live attenuated DENV vaccine candidates. The availability of a panel of well-characterized attenuating mutations that could be introduced into wild type DENV or into incompletely attenuated vaccine candidates to decrease their reactogenicity would promote the development of safe and immunogenic vaccines. Toward this end, we have generated a large panel of mutations in rDEN4 using two approaches: paired charge-to-alanine mutagenesis and chemical mutagenesis.

Paired charge-to-alanine mutagenesis of the DENV-4 NS5 polymerase gene was conducted by mutating the eighty pairs of charged amino acids present in DENV-4 NS5 to Ala-Ala and individually incorporating them into the full length DENV-4 cDNA clone (Hanley et al. 2002). Of the eighty cDNA clones, 32 were found to yield infectious virus upon transfection of Vero and/or C6/36 mosquito cells. In a separate strategy, wild type DENV-4 was grown in Vero cells in the presence of 5-fluorouracil, a chemical mutagen, which resulted in a 100-fold reduction in virus yield (Blaney et al. 2001). Virus progeny were terminally diluted to isolate 1,248 virus clones that were then expanded into small stocks for subsequent screening. Viruses with desirable phenotypes, as described below, were then fully sequenced and the identified mutations were cloned individually into rDEN4 to generate recombinant viruses. Recombinant viruses bearing individual mutations were then studied to confirm that the mutation was responsible for the observed phenotypes. This extra step was not required for the charge-to-alanine mutant viruses because these viruses already contained genetically isolated mutations.

3.2   Selection of Mutant rDEN4 Viruses

The mutant viruses described above were initially screened for temperature sensitivity (*ts*) and small plaque (*sp*) size in Vero cells and the human hepatoma cell line, HuH-7, in an effort to identify presumptive liver-specific attenuating mutations. All mutant viruses that possessed *ts* and/or *sp* phenotypes were also screened for reduced replication in suckling mouse brain as an additional measure of overall level of attenuation. Numerous attenuating mutations conferring an array of the aforementioned attenuation phenotypes were identified and have been described (Blaney et al. 2001, 2002, 2003a; Hanley et al. 2002).

A novel small animal model, SCID-HuH-7 mice, was next developed to identify mutations that might be useful to reduce the low level of hepatotoxicity observed in the rDEN4Δ30 vaccinees. In this animal model of DENV infection, HuH-7 cells are engrafted in SCID mice and allowed to form intraperitoneal tumors into which virus can then be directly injected (Blaney et al. 2002). Subsequent viremia in the SCID-HuH-7 mouse is believed to arise primarily from the human liver cells comprising the tumor mass since DENV administered peripherally to age-matched, nonengrafted mice are highly restricted in replication. Importantly, the relative level of virus replication in SCID-HuH-7 mice correlates with that of rhesus monkeys, which serve as the most widely used animal model for evaluation of the level of attenuation of DENV vaccine candidates (Blaney et al. 2006). In addition, wild type DENV viremia in SCID-HuH-7 mice reaches high levels ($>10^6$ PFU/ml) similar to those observed in humans with severe disease (Vaughn et al. 2000).

Nineteen chemically mutagenized viruses and thirteen charge-to-alanine mutant viruses with *ts* phenotypes in HuH-7 cells were screened in SCID-HuH-7 mice and several were identified with significant reductions in replication compared to wild type DENV-4 (Blaney et al. 2002; Hanley et al. 2004). One such mutation was the 200,201 mutation named for the mutated amino acid pair in the NS5 protein. The 200,201 mutation was found to confer greater than a 100-fold reduction in replication in SCID-HuH-7 mice and importantly, each amino acid change was found to confer this attenuation when the paired mutations were studied separately (Fig. 1). Since both NS5 200 and 201 mutations independently contribute to the attenuation phenotype of rDEN4-200,201, a virus with the pair of mutations should be phenotypically stable since the two mutated codons would require two nucleotide substitutions each to change the alanine substitution to the original charged amino acid residue in the wild type virus (Hanley et al. 2004). The 200,201 mutation and the previously described Δ30 mutation were next combined into a single virus to create the rDEN4Δ30-200,201 vaccine candidate.

Several other mutations that restricted replication in SCID-HuH-7 mice were identified and modified rDEN4Δ30 derivatives were generated with these additional mutations and were shown to be further attenuated in SCID-HuH-7 mice (Hanley et al. 2004). One such virus, rDEN4Δ30-4995, contains the 4995 mutation that was previously found to confer a *ts* phenotype in Vero and HuH-7 cells, to enhance replication in Vero cells at permissive temperature and to cause a 1,000-fold reduction in replication in suckling mouse brain (Blaney et al. 2001; 2003b). Despite the fact that the 4995 mutation alone did not confer a significantly decreased level of replication in SCID-HuH-7 mice, it was discovered that combining the 4995 and Δ30 mutations resulted in a virus with a level of replication in SCID-HuH-7 mice similar to that of rDEN4-200,201 (Fig. 1) (Hanley et al. 2004). In rDEN4Δ30-4995, the mutations at genome position 4995, 4996 and 4997 result in a serine to leucine change at amino acid position 158 of NS3. The mutations incorporated at codon 158 (wt. TCA → mut. CTT) in rDEN4Δ30-4995, would require two nucleotide changes for reversion to wild type sequence. While other modified rDEN4Δ30 viruses have been studied (Hanley et al. 2004), we focus now on rDEN4Δ30-200,201 and rDEN4Δ30-4995, which were found to replicate efficiently in Vero cells and which were attenuated in SCID-HuH-7 mice.

3.3   Preclinical Studies of rDEN4Δ30-200,201 and rDEN4Δ30-4995

The rDEN4Δ30-200,201 and rDEN4Δ30-4995 viruses were evaluated for replication in SCID-HuH-7 mice and found to be at least 250-fold restricted compared to wild type DENV-4 (Fig. 1). These two modified rDEN4Δ30 vaccine candidates were also significantly restricted in replication (40–50-fold) when compared to the rDEN4Δ30 parent virus. Thus, in the SCID-HuH-7 mouse model, both rDEN4Δ30-200,201 and rDEN4Δ30-4995 had clear evidence of further attenuation. The level of attenuation achieved by the combination of the 200,201 and Δ30 mutations was not

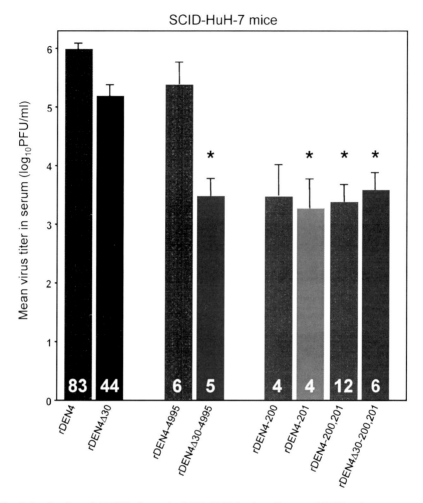

Fig. 1 Replication of rDEN4 viruses in SCID-HuH-7 mice. Groups of SCID mice were transplanted intraperitoneally with 10^7 HuH-7 cells. Approximately six weeks later, tumor-bearing mice received virus injections of 10^4 PFU directly into the tumor. Serum was collected on day 7 and virus titer in serum was determined by plaque assay in Vero cells. Standard error bars and animal numbers in each group are indicated. Asterisks indicate that mean virus titers were significantly lower than rDEN4Δ30 as determined by Tukey-Kramer posthoc test ($P < 0.05$)

greater than the 200,201 mutation alone indicating that the effect of each mutation was not additive. In contrast, the combination of the 4995 and Δ30 mutations synergistically increased the level of attenuation in the SCID-HuH-7 mice.

Mutant viruses rDEN4Δ30-200,201 and rDEN4Δ30-4995 were more attenuated than rDEN4Δ30 in rhesus monkeys (Table 2) and these observations were similar to those in SCID-HuH-7 mice. None of four monkeys inoculated with rDEN4Δ30-200,201 developed detectable viremia, while each of four rDEN4Δ30-vaccinated monkeys was viremic, indicating that the 200,201 mutation restricted

Table 2 Viremia and immunogenicity in rhesus monkeys

Virus[a]	No. of monkeys	% with viremia	Mean no. days with viremia	Mean peak virus titer ($\log_{10}$ PFU/ml)[b]	Geometric mean serum neutralizing antibody titer[c]
rDEN4	8	100	3.0	2.2 ± 0.2	322
rDEN4Δ30	4	100	2.0	1.4 ± 0.2	154
rDEN4Δ30-4995	4	50	0.8	1.1 ± 0.1	79
rDEN4Δ30-200,201	4	0	0.0	< 1.0	139

[a]Groups of rhesus monkeys were inoculated subcutaneously with 10^5 PFU of the indicated virus in a 1 ml dose. Serum was collected on days 0–8, 10, and 28
[b]Virus titer in serum was determined by plaque assay in Vero cells. The limit of detection was 1.0 $\log_{10}$ PFU/ml
[c]Reciprocal plaque reduction (60%) neutralizing antibody titer on day 28

replication of rDEN4Δ30. Despite the lack of detectable viremia, each monkey was infected with rDEN4Δ30-200,201 as demonstrated by a 100% frequency of sero-conversion and complete protection from challenge with wild type DENV-4 (data not shown). Importantly, the GMT was comparable to that of rDEN4Δ30 indicating that rDEN4Δ30-200,201 was not over-attenuated. rDEN4Δ30-4995 had a replication phenotype in rhesus monkeys that appeared to be intermediate between that of rDEN4Δ30 and rDEN4Δ30-200,201 (Table 2). The incidence (50%) and duration (0.8 days) of viremia was approximately half of that observed for rDEN4Δ30, even though the peak virus titer was not different. As observed for monkeys receiving rDEN4Δ30-200,201, all rDEN4Δ30-4995 vaccinated monkeys seroconverted and were completely protected from challenge.

The further attenuation of rDEN4Δ30 in SCID-HuH-7 mice and rhesus monkeys by the introduction of the 200,201 or 4995 mutation indicated that these two modified vaccine candidates should be evaluated in humans. Such a study would not only allow for evaluation of clinical reactogenicity but would also be useful in determining if the differences in replication between these vaccine candidates and their rDEN4Δ30 parent virus that were seen in both SCID-HuH-7 mice and rhesus monkeys would also be seen in humans. Such a correlation would serve as a guide for the use of these preclinical phenotypes in making decisions about which vaccine candidates should be evaluated in human subjects. Because the rDEN4Δ30-200,201 vaccine candidate was the most attenuated in rhesus monkeys and has the added advantage of increased genetic stability, it was chosen first to be tested in humans as described below. Clinical evaluation of rDEN4Δ30-4995 is currently ongoing.

4   Clinical Evaluation of rDEN4Δ30-200,201

The rDEN4Δ30-200,201 vaccine candidate was evaluated at a dose of 10^5 PFU in a double-blind phase I study of 28 volunteers including 8 placebo recipients and compared to the response of volunteers in the previous study of rDEN4Δ30 given at a 10^5 PFU dose (McArthur et al. 2008). As expected, volunteers receiving the rDEN4Δ30-200,201 vaccine did not develop significant local or systemic illness. Viremia was not detectable in any vaccinee given rDEN4Δ30-200,201, which is a striking difference from the 70% incidence of viremia that was observed in the previous study of rDEN4Δ30 at the 10^5 PFU dose (Table 3). Even if a low level of viremia for rDEN4Δ30-200,201 was present below the limit of detection (0.5 $\log_{10}$ PFU/ml), it would be at least tenfold reduced compared to that of rDEN4Δ30 (1.6 $\log_{10}$ PFU/ml). Despite the lack of detectable viremia in rDEN4Δ30-200,201 vaccinees, all volunteers were infected as indicated by the 100% seroconversion rate on day 28 (data not shown). On day 42, the neutralizing antibody response against DENV-4 induced by rDEN4Δ30-200,201 (GMT of 1:84) was less than twofold different than that observed for rDEN4Δ30 (GMT of 1:156) (Table 3). This suggests that sufficient replication of rDEN4Δ30-200,201 occurred in the vaccinees to result in the induction of a moderate level of immunity. The virological and immunological results of this study correlate well with the observations seen in the rhesus monkey studies of rDEN4Δ30-200,201 (Table 2 vs. Table 3) confirming the utility of this animal model in the preclinical evaluation of DENV-4 vaccines.

The reactogenicity profile of rDEN4Δ30-200,201 was also addressed in this study (McArthur et al. 2008). The incidence of rash was lower in the rDEN4Δ30-200,201 vaccinees (20%) when compared to the rDEN4Δ30 vaccinees (50%) at the 10^5 PFU dose. However, the most conspicuous difference between the cohorts was the lack of ALT elevations in volunteers receiving rDEN4Δ30-200,201 versus the 25% incidence in the rDEN4Δ30 vaccinees. This observation indicates that the mild hepatotoxicity of rDEN4Δ30 was abrogated by the inclusion of the 200,201 mutation and suggests that the 200,201 mutation restricted replication specifically in the liver of vaccinees. It is possible that the 200,201 mutation also decreased replication at a peripheral site.

These studies indicate that the selection of mutations to further attenuate rDEN4Δ30 for SCID-HuH-7 mice was a rational approach to vaccine design and resulted in the intended decrease in the level of reactogenicity of rDEN4Δ30-200,201 for humans while maintaining sufficient immunogenicity. In the current case, a mutation that restricted replication in HuH-7 cells, a human liver cell line, also decreased the hepatotoxicity of rDEN4Δ30 for humans. This approach has also been applied to the development of attenuated mutant neurotropic flaviviruses that are selected for restricted replication in neuronal cells in culture and that were subsequently found to have decreased neurovirulence (Rumyantsev et al. 2006). The strong immunogenicity of rDEN4Δ30-200,201, despite its reduced viremia and putative restricted replication in the liver compared to rDEN4Δ30, suggests that this vaccine candidate is replicating efficiently in some tissue(s) of the host. Using reverse genetics to develop these vaccine candidates has permitted the rapid importation of desired attenuating mutations into wild type virus or incompletely attenuated mutants. The success of this study confirms the utility of such an approach to live attenuated vaccine development in which mutations are added sequentially to a virus with clinical testing after each addition to identify a vaccine

Table 3 rDEN4Δ30-200,201 is more attenuated than rDEN4Δ30 when administered as a 10^5 PFU dose in 20 volunteers

Virus	% of volunteers with viremia	Mean peak virus titer ± SE ($\log_{10}$ PFU/ml serum)[a]	Geometric mean serum neutralizing antibody titer[b]	% seroconversion[c]	% of volunteers with indicated clinical sign:				
					Fever	Rash	Headache	Neutropenia[d]	Elevated ALT level[e]
rDEN4Δ30[f]	70	1.6 ± 0.1	156	100	5[g]	50	35	15	25
rDEN4Δ300-200,201	0	< 0.5	84	100[h]	5[i]	20	30	10	0[j]
Placebo[k]	0	n/a	<10	0	0	0	25	25	0

[a]Calculated for viremic volunteers only. The lower limit of detection is 0.5 $\log_{10}$ PFU/ml serum

[b]Reciprocal plaque reduction (60%) neutralizing antibody titer of day 42 serum

[c]Percent seroconversion defined as a fourfold or greater increase in serum neutralizing antibody level to DENV-4 on day 28 or 42

[d]Neutropenia defined as an absolute neutrophil count of ≤ 1,500 cells/mm^3

[e]Elevated ALT level defined as any value above the upper limit of normal (for males, >72 U/L^{-1}; for females, >52 U/L^{-1})

[f]Historical data, except for the neutralizing antibody titer that was performed concurrently with the rDEN4Δ30-200,201 group. Although the neutralizing antibody data shown in Tables 1 and 3 are for the same sera from the rDEN4Δ30 group, the mean values differ slightly, reflecting assay to assay variability

[g]Fever occurred on days 3 and 5; maximum temp was 100.5°F

[h]One volunteer was excluded from serology because of pregnancy; total of 19 volunteers tested

[i]Fever occurred on day 15; maximum temp was 100.6°F

[j]Incidence significantly lower than rDEN4Δ30 group (25%) determined by Fisher Exact Test (P < 0.05)

[k]n = 8

virus that has achieved an acceptable balance between attenuation and immunogenicity. Based on the highly restricted replication, lower reactogenicity and similar immunogenicity of rDEN4Δ30-200,201 compared to rDEN4Δ30, the new vaccine candidate can be strongly considered for inclusion in a tetravalent vaccine formulation. At present we favor the use of the rDEN4Δ30 at a dose of 10^3 PFU since it is safe, economical to manufacture and immunogenic at this dose, with rDEN4Δ30-200,201 serving as a backup vaccine candidate, if needed. In addition, the 200,201 mutation may be also useful for further attenuating the DENV, West Nile virus and tick-borne encephalitis virus antigenic chimeric viruses that have been created using the rDEN4Δ30 virus (Durbin et al. 2006b; Pletnev et al. 2001, 2006).

5 Future Directions

Current studies in humans focus on the analysis of safety, infectivity and immunogenicity of individual vaccine candidates, including rDEN4Δ30-200,201 and rDEN4Δ30-4995, so that a tetravalent formulation can be identified. The development and clinical evaluation of rDEN4Δ30 has led to two strategies for development of vaccine candidates for the other DENV serotypes (Blaney et al. 2006). First, reverse genetics has been used to introduce the structurally conserved attenuating Δ30 mutation into the 3′-UTR of cDNA clones of DENV-1, DENV-2 and DENV-3 (Blaney et al. 2004a, b, 2005; Durbin et al. 2006a; Whitehead et al. 2003a). Alternatively, antigenic chimeric viruses have been generated by replacement of the structural proteins of the attenuated rDEN4Δ30 vaccine candidate with those from DENV-1, DENV-2 or DENV-3 (Blaney et al. 2004a; 2007, Durbin et al. 2006b; Whitehead et al. 2003b). Targeted mutagenesis of the DENV-3 3′-UTR has also been used to identify additional attenuated vaccine candidates for this serotype (Blaney et al. 2008). In conclusion, the ability to generate and modify vaccine candidates by reverse genetics and targeted mutagenesis has increased the likelihood that four suitable vaccine candidates can be identified for a tetravalent formulation. However, additional research is required for the identification of a suitable tetravalent DENV vaccine and for the implementation of its use.

Acknowledgments These studies were supported by the National Institute of Allergy and Infectious Diseases Division of Intramural Research. We thank Kathryn Hanley, Cai-Yen Firestone and Christopher Hanson for their contributions to these studies.

References

Bhamarapravati N, Yoksan S (1997) Live attenuated tetravalent dengue vaccine. In: Gubler DJ, Kuno G (eds) Dengue and dengue hemorrhagic fever. CAB International, New York, pp 367–377
Blaney JE Jr, Durbin AP, Murphy BR, Whitehead SS (2006) Viral Immunol 19:10–32

Blaney JE Jr, Hanson CT, Firestone CY, Hanley KA, Murphy BR, Whitehead SS (2004a) Am J Trop Med Hyg 71:811–821

Blaney JE Jr, Hanson CT, Hanley KA, Murphy BR, Whitehead SS (2004b) BMC Infect Dis 4:39

Blaney JE Jr, Johnson DH, Firestone CY, Hanson CT, Murphy BR, Whitehead SS (2001) J Virol 75:9731–9740

Blaney JE Jr, Johnson DH, Manipon GG, Firestone CY, Hanson CT, Murphy BR, Whitehead SS (2002) Virology 300:125–139

Blaney JE Jr, Manipon GG, Murphy BR, Whitehead SS (2003a) Arch Virol 148:999–1006

Blaney JE Jr, Matro JM, Murphy BR, Whitehead SS (2005) J Virol 79:5516–5528

Blaney JE Jr, Sathe NS, Goddard L, Hanson CT, Romero TA, Hanley KA, Murphy BR, Whitehead SS (2008) Vaccine 26:817–828

Blaney JE Jr, Sathe NS, Hanson CT, Firestone CY, Murphy BR, Whitehead SS (2007) Virology Journal 4:23

Blaney JE, Manipon GG, Firestone CY, Johnson DH, Hanson CT, Murphy BR, Whitehead SS (2003b) Vaccine 21:4317–4327

Durbin AP, Karron RA, Sun W, Vaughn DW, Reynolds MJ, Perreault JR, Thumar B, Men R, Lai CJ, Elkins WR, Chanock RM, Murphy BR, Whitehead SS (2001) Am J Trop Med Hyg 65:405–413

Durbin AP, McArthur J, Marron JA, Blaney JE Jr, Thumar B, Wanionek K, Murphy BR, Whitehead SS (2006a) Human Vaccines 2:167–173

Durbin AP, McArthur JH, Marron JA, Blaney JE, Thumar B, Wanionek K, Murphy BR, Whitehead SS (2006b) Human Vaccines 2:255–260

Durbin AP, Whitehead SS, McArthur J, Perreault JR, Blaney JE Jr, Thumar B, Murphy BR, Karron RA (2005) rDEN4 Delta 30, a live attenuated dengue virus type 4 vaccine candidate, is safe, immunogenic, and highly infectious in healthy adult volunteers. J Infect Dis 191:710–718

Eckels KH, Scott RM, Bancroft WH, Brown J, Dubois DR, Summers PL, Russell PK, Halstead SB (1984) Am J Trop Med Hyg 33:684–689

Edelman R, Tacket CO, Wasserman SS, Vaughn DW, Eckels KH, Dubois DR, Summers PL, Hoke CH (1994) J Infect Dis 170:1448–1455

Endy TP, Nisalak A, Chunsuttitwat S, Vaughn DW, Green S, Ennis FA, Rothman AL, Libraty DH (2004) J Infect Dis 189:990–1000

Gubler DJ (1997) Dengue and dengue hemorrhagic fever: its history and resurgence as a global public health problem. In: Gubler DJ, Kuno G (eds) Dengue and dengue hemorrhagic fever. CAB International, New York, pp 1–22

Gubler DJ, Meltzer M (1999) Adv Virus Res 53:35–70

Guirakhoo F, Kitchener S, Morrison D, Forrat R, McCarthy K, Nichols R, Yoksan S, Duan X, Ermak TH, Kanesa-Thasan N, Bedford P, Lang J, Quentin-Millet MJ, Monath TP (2006) Human Vaccines 2:60–67

Halstead SB (2003) Adv Virus Res 60:421–467

Halstead SB, Marchette NJ (2003) Am J Trop Med Hyg 69:5–11

Hanley KA, Lee JJ, Blaney JE Jr, Murphy BR, Whitehead SS (2002) J Virol 76:525–531

Hanley KA, Manlucu LR, Manipon GG, Hanson CT, Whitehead SS, Murphy BR, Blaney JE Jr (2004) Vaccine 22:3440–3448

Kalayanarooj S, Vaughn DW, Nimmannitya S, Green S, Suntayakorn S, Kunentrasai N, Viramitrachai W, Ratanachu-eke S, Kiatpolpoj S, Innis BL, Rothman AL, Nisalak A, Ennis FA (1997) J Infect Dis 176:313–321

Kanesa-thasan N, Sun W, Kim-Ahn G, Van Albert S, Putnak JR, King A, Raengsakulrach B, Christ-Schmidt H, Gilson K, Zahradnik JM, Vaughn DW, Innis BL, Saluzzo JF, Hoke CH Jr (2001) Vaccine 19:3179–3188

Kitchener S, Nissen M, Nasveld P, Forrat R, Yoksan S, Lang J, Saluzzo JF (2006) Vaccine 24:1238–1241

Kuo CH, Tai DI, Chang-Chien CS, Lan CK, Chiou SS, Liaw YF (1992) Am J Trop Med Hyg 47:265–270

Lai CJ, Zhao BT, Hori H, Bray M (1991) Proc Natl Acad Sci USA 88:5139–5143

McArthur JH, Durbin AP, Marron JA, Wanionek KA, Thumar B, Pierro DI, Schmidt AC, Blaney JE Jr, Murphy BR, Whitehead SS (2008) Phase I Clinical Evaluation of rDEN4{Delta}30-200, 201: A Live Attenuated Dengue 4 Vaccine Candidate Designed for Decreased Hepatotoxicity. Am J Trop Med Hyg 79:678–684

Men R, Bray M, Clark D, Chanock RM, Lai CJ (1996) J Virol 70:3930–3937

Mohan B, Patwari AK, Anand VK (2000) J Trop Pediatr 46:40–43

Pletnev AG, Bray M, Hanley KA, Speicher J, Elkins R (2001) J Virol 75:8259–8267

Pletnev AG, Swayne DE, Speicher J, Rumyantsev AA, Murphy BR (2006) Vaccine 24:6392–6404

Rice CM, Grakoui A, Galler R, Chambers TJ (1989) New Biol 1:285–296

Rumyantsev AA, Murphy BR, Pletnev AG (2006) Journal of virology 80:1427–1439

Sabchareon A, Lang J, Chanthavanich P, Yoksan S, Forrat R, Attanath P, Sirivichayakul C, Pengsaa K, Pojjaroen-Anant C, Chambonneau L, Saluzzo JF, Bhamarapravati N (2004) Pediatr Infect Dis J 23:99–109

Sabchareon A, Lang J, Chanthavanich P, Yoksan S, Forrat R, Attanath P, Sirivichayakul C, Pengsaa K, Pojjaroen-Anant C, Chokejindachai W, Jagsudee A, Saluzzo JF, Bhamarapravati N (2002) Am J Trop Med Hyg 66:264–272

Sabin AB (1952) Am J Trop Med Hyg 1:30–50

Troyer JM, Hanley KA, Whitehead SS, Strickman D, Karron RA, Durbin AP, Murphy BR (2001) Am J Trop Med Hyg 65:414–419

Vaughn DW, Green S, Kalayanarooj S, Innis BL, Nimmannitya S, Suntayakorn S, Endy TP, Raengsakulrach B, Rothman AL, Ennis FA, Nisalak A (2000) J Infect Dis 181:2–9

Vaughn DW, Hoke CH Jr, Yoksan S, LaChance R, Innis BL, Rice RM, Bhamarapravati N (1996) Vaccine 14:329–336

Wahid SF, Sanusi S, Zawawi MM, Ali RA (2000) Southeast Asian J Trop Med Public Health 31:259–263

Whitehead SS, Blaney JE, Durbin AP, Murphy BR (2007) Nature Rev. Microbiol 5:518–528

Whitehead SS, Falgout B, Hanley KA, Blaney JE Jr, Markoff L, Murphy BR (2003a) J Virol 77:1653–1657

Whitehead SS, Hanley KA, Blaney JE, Gilmore LE, Elkins WR, Murphy BR (2003b) Vaccine 21:4307–4316

WHO (1997) Dengue haemorrhagic fever: diagnosis, treatment prevention and control. WHO, Geneva

Index